科技大讲堂丛书

Oracle Database System Management and Maintenance

Oracle数据库
系统管理与运维
微课视频版

张立杰 ◎ 主编
Zhang Lijie

陈恒　陶永鹏 ◎ 副主编
Chen Heng　Tao Yongpeng

清华大学出版社
北京

内 容 简 介

本书以 Oracle 19c 为平台编写,全书分为 3 部分,共 15 章。第 1 部分数据库系统,包括第 1~4 章,Oracle 体系结构、Oracle 实例、Oracle 数据库、数据字典与动态性能视图;第 2 部分数据库管理,包括第 5~9 章,Oracle 数据库常用工具、Oracle 空间管理、Oracle 网络配置管理、Oracle 监控管理、数据库的归档模式管理;第 3 部分数据库运维,包括第 10~15 章,Oracle 备份、Oracle 恢复、Oracle 数据的移动、Oracle 闪回技术、Oracle 并发与一致、Oracle 优化。

本书以基础知识点与知识点案例及项目案例相结合的方式,从系统到管理再到运维,由浅入深地为读者展开 Oracle 的讲解,具有内容全面、案例丰富、代码具体等优点。本书每个知识点案例和项目案例都详细地阐述了具体的操作步骤,并添加了实际运行代码。本书还有教学课件、全部源代码和微课视频等丰富的配套资源。

本书适合 Oracle 数据库的学习者和从业者使用,也适合作为开发人员的查阅和参考资料,并且非常适合作为高等院校、培训机构的教材。

本书封面贴有清华大学出版社防伪标签,无标签者不得销售。
版权所有,侵权必究。举报: 010-62782989,beiqinquan@tup.tsinghua.edu.cn。

图书在版编目(CIP)数据

Oracle 数据库系统管理与运维: 微课视频版/张立杰主编. —北京: 清华大学出版社,2021.1(2023.9重印)
(清华科技大讲堂丛书)
ISBN 978-7-302-56610-6

Ⅰ. ①O… Ⅱ. ①张… Ⅲ. ①关系数据库系统 Ⅳ. ①TP311.138

中国版本图书馆 CIP 数据核字(2020)第 192661 号

责任编辑: 陈景辉
封面设计: 刘　键
责任校对: 徐俊伟
责任印制: 丛怀宇

出版发行: 清华大学出版社
　　网　　址: http://www.tup.com.cn, http://www.wqbook.com
　　地　　址: 北京清华大学学研大厦 A 座　　邮　编: 100084
　　社 总 机: 010-83470000　　邮　购: 010-62786544
　　投稿与读者服务: 010-62776969, c-service@tup.tsinghua.edu.cn
　　质量反馈: 010-62772015, zhiliang@tup.tsinghua.edu.cn
　　课件下载: http://www.tup.com.cn,010-83470236
印 装 者: 三河市人民印务有限公司
经　　销: 全国新华书店
开　　本: 185mm×260mm　　印　张: 20　　字　数: 500 千字
版　　次: 2021 年 3 月第 1 版　　印　次: 2023 年 9 月第 3 次印刷
印　　数: 2501~3000
定　　价: 69.90 元

产品编号: 084754-01

前言

无论是一名 IT 领域的从业者还是计算机相关专业的在读学生,无论你在工作和学习中偏重系统理论、程序开发还是平台运维,都不可避免地要与数据库打交道。根据 2020 年 2 月 DB-Engines 发布的数据库流行度排行榜的数据显示,Oracle 数据库依然雄踞第一宝座,由此可见 Oracle 数据库的重要性。

我于 2008 年开始讲解 Oracle 数据库的相关课程,很多企业也邀请我进行 Oracle 相关培训。在此过程中,通过不断地实践与探索,完成了这本内容全面、知识递进、深入浅出的数据库教材。本书全面涵盖数据库系统结构、数据库管理以及数据运维等知识,提供了从基础到管理,再到运维所具备的基本知识,以 Oracle 19c 为平台编写,全书分为 3 部分,共 15 章。第 1 部分数据库系统,包括第 1~4 章,Oracle 体系结构、Oracle 实例、Oracle 数据库、数据字典与动态性能视图;第 2 部分数据库管理,包括第 5~9 章,Oracle 数据库常用工具、Oracle 空间管理、Oracle 网络配置管理、Oracle 监控管理、数据库的归档模式管理;第 3 部分数据库运维,包括第 10~15 章,Oracle 备份、Oracle 恢复、Oracle 数据的移动、Oracle 闪回技术、Oracle 并发与一致性、Oracle 优化。

本书特点

(1) 循序渐进、由浅入深地带领读者进入 Oracle 的世界。

(2) 基础知识点与实战案例相结合,案例丰富,提供脚本。

本书的每个知识点都通过通俗易懂的知识点案例进行讲解与实操,详细地讲述了实际应用中所需的各类知识,并配备案例具体的脚本和运行结果。

(3) 精选案例、步骤清晰、可操作性强。

本书针对每章的重要知识点分别配有一个综合案例。综合案例步骤清晰,代码明晰,以帮助读者可以跟着案例的讲解,一步步地完成操作,达到温故知新、付诸实践的效果。本书所有案例都在 Oracle 19c 平台上测试,因此读者须使用 Oracle 19c 版本。

(4) 配套视频讲解,实现手把手式教学。

本书每章都提供微课视频,手把手地带领读者完成案例操作。

配套资源

为便于教学,本书配有 550 分钟微课视频、源代码、教学课件、教学大纲、教案。

(1) 获取教学视频方式:读者可以先扫描本书封底的文泉云盘防盗码,再扫描书中相应的视频二维码,观看教学视频。

(2) 获取源代码方式:先扫描本书封底的文泉云盘防盗码,再扫描下方二维码,即可获取。

源代码

书中源码均以小写形式输入,正文中出现的语法、参数、表名等均以小写形式呈现,除表空间、值、路径、结果中的引述等。

(3) 其他配套资源可以扫描本书封底的课件二维码下载。

本书适合 Oracle 数据库的学习者和从业者使用,也适合作为开发人员的查阅和参考资料,也非常适合作为高等院校、培训机构的教材。

在此对参与本书编写的同仁们的辛勤付出和出色工作表示衷心的感谢。

由于水平有限,书中难免有不足之处,敬请广大读者批评和指正。

作　者

2021 年 1 月

目　录

第1部分　数据库系统

第1章　Oracle体系结构 ……………………………………………………………………… 3

1.1　Oracle数据库系统的发展史 …………………………………………………………… 3
1.2　Oracle数据库系统的安装 ……………………………………………………………… 4
　　1.2.1　Oracle 19c的下载 ………………………………………………………………… 4
　　1.2.2　Oracle 19c的安装 ………………………………………………………………… 5
　　1.2.3　Oracle 19c安装测试 …………………………………………………………… 10
1.3　Oracle数据库服务器 ………………………………………………………………… 13
1.4　Oracle体系结构 ……………………………………………………………………… 14

第2章　Oracle实例 …………………………………………………………………………… 17

2.1　数据库实例 …………………………………………………………………………… 17
　　2.1.1　内存结构 ……………………………………………………………………… 17
　　2.1.2　进程结构 ……………………………………………………………………… 18
2.2　数据库内存结构 ……………………………………………………………………… 18
　　2.2.1　SGA …………………………………………………………………………… 18
　　2.2.2　PGA …………………………………………………………………………… 23
2.3　数据库进程 …………………………………………………………………………… 24
　　2.3.1　用户进程 ……………………………………………………………………… 24
　　2.3.2　服务器进程 …………………………………………………………………… 25
　　2.3.3　后台进程 ……………………………………………………………………… 26
2.4　进程与内存的交互情况 ……………………………………………………………… 30
2.5　项目案例 ……………………………………………………………………………… 31
　　2.5.1　用户进程实例 ………………………………………………………………… 31
　　2.5.2　服务器进程实例 ……………………………………………………………… 33
　　2.5.3　SGA实例 ……………………………………………………………………… 34

第3章　Oracle数据库 ………………………………………………………………………… 36

3.1　Oracle数据库概述 …………………………………………………………………… 36

3.2 Oracle 数据库物理存储 ………………………………………………………………… 37
　　3.2.1 数据文件 …………………………………………………………………… 38
　　3.2.2 控制文件 …………………………………………………………………… 40
　　3.2.3 日志文件 …………………………………………………………………… 43
　　3.2.4 参数文件 …………………………………………………………………… 46
　　3.2.5 其他文件 …………………………………………………………………… 48
3.3 Oracle 启动与关闭 ……………………………………………………………………… 48
　　3.3.1 Oracle 的启动过程 ………………………………………………………… 49
　　3.3.2 数据库的启动 ……………………………………………………………… 51
　　3.3.3 数据库的关闭过程 ………………………………………………………… 52
　　3.3.4 数据库的关闭 ……………………………………………………………… 52
3.4 Oracle 逻辑存储 ………………………………………………………………………… 54
　　3.4.1 表空间 ……………………………………………………………………… 54
　　3.4.2 段 …………………………………………………………………………… 58
　　3.4.3 区 …………………………………………………………………………… 58
　　3.4.4 块 …………………………………………………………………………… 59
3.5 项目案例 ………………………………………………………………………………… 60
　　3.5.1 内存结构实例 ……………………………………………………………… 60
　　3.5.2 日志文件操作实例 ………………………………………………………… 60
　　3.5.3 参数文件操作实例 ………………………………………………………… 62
　　3.5.4 数据库打开与关闭实例 …………………………………………………… 63

第 4 章　数据字典与动态性能视图 ……………………………………………………… 66

4.1 静态数据字典 …………………………………………………………………………… 66
　　4.1.1 通用概要类型的数据字典视图 …………………………………………… 67
　　4.1.2 常用的数据字典视图 ……………………………………………………… 67
4.2 动态性能视图 …………………………………………………………………………… 69
4.3 项目案例 ………………………………………………………………………………… 71

第 2 部分　数据库管理

第 5 章　Oracle 数据库常用工具 ………………………………………………………… 77

5.1 数据库开发工具 ………………………………………………………………………… 77
5.2 数据库开发人员的主题 ………………………………………………………………… 79
5.3 数据库管理工具 ………………………………………………………………………… 79
5.4 数据库管理人员的主题 ………………………………………………………………… 82
5.5 项目案例 ………………………………………………………………………………… 82

第 6 章 Oracle 空间管理 …… 93

- 6.1 数据库的创建与配置 …… 94
- 6.2 表空间的管理 …… 94
  - 6.2.1 表空间的管理方式 …… 95
  - 6.2.2 表空间管理 …… 96
  - 6.2.3 还原表空间 …… 99
  - 6.2.4 临时表空间 …… 101
  - 6.2.5 用户与表空间 …… 104
  - 6.2.6 表空间监控与注意事项 …… 104
- 6.3 段的管理 …… 105
- 6.4 区的管理 …… 106
- 6.5 块的管理 …… 106
- 6.6 项目案例 …… 110
  - 6.6.1 对表空间相关的操作实例 …… 110
  - 6.6.2 用户与表空间操作实例 …… 112
  - 6.6.3 表段与块操作实例 …… 113
  - 6.6.4 数据库行链操作实例 …… 118
  - 6.6.5 数据库行迁移操作实例 …… 119

第 7 章 Oracle 网络配置管理 …… 124

- 7.1 Oracle 数据库网络配置 …… 124
  - 7.1.1 Oracle 网络服务组件 …… 124
  - 7.1.2 Oracle 网络连接的流程 …… 125
  - 7.1.3 Oracle 网络配置文件 …… 125
- 7.2 网络概要配置 …… 126
  - 7.2.1 网络概要配置步骤 …… 126
  - 7.2.2 网络概要配置文件——sqlnet.ora …… 128
- 7.3 服务器端网络配置——监听器 …… 129
  - 7.3.1 监听器的配置 …… 129
  - 7.3.2 监听配置文件——listener.ora …… 133
  - 7.3.3 监听器的管理 …… 134
- 7.4 客户端网络配置——网络服务名 …… 136
  - 7.4.1 网络服务名配置 …… 136
  - 7.4.2 网络服务名配置文件——tnsnames.ora …… 140
- 7.5 网络连接 …… 140
- 7.6 服务的启动和停止 …… 142
- 7.7 项目案例 …… 143

第 8 章 Oracle 监控管理 … 146

8.1 自动诊断知识库 … 146
8.2 告警 … 149
8.3 跟踪 … 149
8.3.1 后台进程跟踪文件 … 149
8.3.2 用户进程跟踪文件 … 150
8.4 审计 … 153
8.4.1 审计概述 … 153
8.4.2 统一审计 … 154
8.4.3 审计策略 … 155
8.4.4 审计配置 … 156
8.4.5 审计的使用 … 158
8.5 项目案例 … 158
8.5.1 审计策略实例 … 158
8.5.2 语句审计实例 … 160
8.5.3 权限审计实例 … 161
8.5.4 对象审计实例 … 163

第 9 章 数据库的归档模式管理 … 165

9.1 非归档模式与归档模式 … 165
9.2 归档模式设置 … 167
9.2.1 归档模式的查询 … 167
9.2.2 归档模式间切换 … 168
9.2.3 归档模式设置 … 168
9.3 归档重做日志 … 170
9.3.1 归档日志切换 … 170
9.3.2 归档进程 … 171
9.3.3 归档日志文件和归档目录 … 172
9.4 归档模式下相关数据视图和脚本 … 174
9.5 项目案例 … 177
9.5.1 归档模式切换实例 … 177
9.5.2 归档日志文件实例 … 178

第 3 部分 数据库运维

第 10 章 Oracle 备份 … 183

10.1 数据库故障 … 183
10.2 数据库备份概述 … 185

    10.2.1 冷备份 ……………………………………………………………… 185

    10.2.2 热备份 ……………………………………………………………… 187

  10.3 项目案例 ………………………………………………………………………… 188

    10.3.1 冷备份实例 ………………………………………………………… 188

    10.3.2 热备份实例 ………………………………………………………… 190

    10.3.3 冷备份自动化实例 ………………………………………………… 193

    10.3.4 热备份自动化实例 ………………………………………………… 194

第 11 章 Oracle 恢复 …………………………………………………………………… 196

  11.1 数据库恢复概述 ………………………………………………………………… 196

    11.1.1 实例恢复 …………………………………………………………… 197

    11.1.2 介质恢复 …………………………………………………………… 197

  11.2 非归档模式下的数据库恢复 …………………………………………………… 198

  11.3 归档模式下的数据库完全恢复 ………………………………………………… 198

    11.3.1 概念 ………………………………………………………………… 198

    11.3.2 恢复的级别 ………………………………………………………… 198

    11.3.3 归档模式下数据库完全恢复 ……………………………………… 201

  11.4 有归档日志的数据库不完全恢复 ……………………………………………… 201

    11.4.1 数据库不完全恢复概述 …………………………………………… 201

    11.4.2 数据文件损坏的数据库不完全恢复 ……………………………… 202

    11.4.3 控制文件损坏的数据库不完全恢复 ……………………………… 203

    11.4.4 日志文件损坏的数据库不完全恢复 ……………………………… 203

  11.5 项目案例 ………………………………………………………………………… 204

    11.5.1 非归档模式下的备份与完全恢复实例 …………………………… 204

    11.5.2 联机备份下数据文件损毁的恢复实例 …………………………… 209

    11.5.3 损坏日志文件的恢复实例 ………………………………………… 218

    11.5.4 损坏控制文件的恢复实例 ………………………………………… 221

    11.5.5 基于时间点的不完全恢复实例 …………………………………… 225

    11.5.6 基于 SCN 的不完全恢复实例 …………………………………… 229

第 12 章 Oracle 数据的移动 ……………………………………………………………… 230

  12.1 逻辑备份与逻辑恢复 …………………………………………………………… 230

  12.2 数据泵技术 ……………………………………………………………………… 230

  12.3 数据泵导出 ……………………………………………………………………… 231

    12.3.1 EXPDP 导出概述 ………………………………………………… 231

    12.3.2 EXPDP 参数介绍 ………………………………………………… 232

    12.3.3 EXPDP 导出实例 ………………………………………………… 233

  12.4 数据泵导入 ……………………………………………………………………… 236

    12.4.1 IMPDP 导入概述 ………………………………………………… 236

12.4.2 IMPDP 参数介绍 236
12.4.3 IMPDP 导入实例 237
12.5 项目案例 239
12.5.1 逻辑备份与逻辑恢复实例 239
12.5.2 数据移动实例 243

第 13 章 Oracle 闪回技术 245

13.1 闪回技术 245
13.2 闪回设置 245
13.3 闪回查询 247
13.4 闪回版本查询 247
13.5 闪回表 248
13.6 闪回删除 248
13.7 闪回数据库 250
13.8 项目案例 250
13.8.1 闪回查询实例 250
13.8.2 闪回删除实例 255
13.8.3 闪回表实例 259
13.8.4 闪回数据库实例 261

第 14 章 Oracle 并发与一致性 266

14.1 并发性与一致性 266
14.2 隔离机制 268
14.3 锁机制 269
14.4 项目案例 271
14.4.1 Read Committed 隔离级别实例 271
14.4.2 Serializable 事务隔离级别实例 272
14.4.3 提交读与序列实例 274
14.4.4 行锁定实例 276

第 15 章 Oracle 优化 279

15.1 Oracle 性能优化概述 279
15.2 系统设计优化 281
15.3 内存优化 283
15.3.1 SGA 内存优化 283
15.3.2 PGA 内存优化 287
15.4 I/O 优化 289
15.4.1 Oracle 中 I/O 的产生 289
15.4.2 I/O 竞争文件与数据的优化 290

15.4.3 I/O 物理文件的优化 …………………………………………………… 291
15.4.4 I/O 调优的其他手段 …………………………………………………… 294
15.5 SQL 语句优化 ……………………………………………………………………… 295
15.5.1 SQL 语句优化原则 …………………………………………………… 295
15.5.2 SQL 语句执行计划 …………………………………………………… 296
15.6 项目案例 …………………………………………………………………………… 299
15.6.1 SQL 语句跟踪与优化实例 …………………………………………… 299
15.6.2 缓存命中率实例 ……………………………………………………… 304

第 1 部分

数据库系统

第 1 章　Oracle 体系结构

第 2 章　Oracle 实例

第 3 章　Oracle 数据库

第 4 章　数据字典与动态性能视图

第1章 Oracle体系结构

PPT 视频讲解

1.1 Oracle 数据库系统的发展史

Oracle 是甲骨文公司的一款关系数据库管理系统,也是目前较流行的关系数据库管理系统,在数据库领域中一直处于领先地位。该系统具有可移植性好、使用方便、功能强等特点。

Oracle 数据库自发布至今,也经历了一个从不稳定到稳定,从功能简单至强大的过程。从第 2 版开始,Oracle 的每一次版本升级,都具有里程碑意义。

(1) 1979 年的夏季,RSI 公司(Relational Software,Inc.,Oracle 公司的前身)发布了 Oracle 第 2 版。

(2) 1983 年 3 月,RSI 公司发布了 Oracle 第 3 版,其有了一个关键的特性——可移植性。

(3) 1984 年 10 月,Oracle 公司(RSI 公司更名为 Oracle)发布了第 4 版产品,其增加了读一致性这个重要特性。

(4) 1985 年,Oracle 公司发布了 5.0 版。这个版本是比较稳定的版本,并实现了 C/S 模式工作。

(5) 1986 年,Oracle 公司发布了 5.1 版。该版本开始支持分布式查询。

(6) 1988 年,Oracle 公司发布了第 6 版。该版本中引入了行级锁特性,同时还引入了联机热备份功能。

(7) 1992 年 6 月,Oracle 公司发布了第 7 版。该版本增加了包括分布式事务处理、用于应用程序开发的新工具及安全性方法等功能。

(8) 1997 年 6 月,Oracle 公司发布了第 8 版。Oracle 8 支持面向对象的开发及新的多媒体应用。

(9) 1998 年 9 月,Oracle 公司正式发布 Oracle 8i。正是因为该版本对 Internet 的支持,所以在版本号后添加了标志 i,代表 Internet。

(10) 2001 年 6 月，Oracle 公司发布了 Oracle 9i。

(11) 2003 年 9 月，Oracle 公司发布了 Oracle 10g。这一版的最大特性就是加入了网格计算的功能，因此版本号后的标志使用了字母 g，代表 Grid 网格。

(12) 2007 年 7 月 11 日，Oracle 公司发布了 Oracle 11g。Oracle 11g 实现了信息生命周期管理(Information Lifecycle Management)等多项创新。

(13) 2013 年 6 月 26 日，Oracle 公司发布了 Oracle 12c。该版本引入了 CDB 与 PDB 的新特性，在 Oracle 12c 数据库引入的多租用户环境中，允许一个数据库容器(CDB)承载多个可插拔数据库(PDB)。在 Oracle 12c 之前，实例与数据库是一对一或多对一关系(RAC)。当进入 Oracle 12c 后，实例与数据库可以是一对多的关系。

(14) 2018 年 2 月 16 日发布的 Oracle 18c 继续秉承了 Oracle 的 Cloud First 理念，在 Cloud 和 Engineered Systems 上推出。Oracle 18c 号称是一款自治性数据库，可以减少很多 DBA 的工作量。

(15) 2019 年 2 月发布了 Oracle Database 19c(以下统称为 Oracle 19c)，也就是 12.2.0.3 将是 Oracle 12c 的终极版本，相当于传统的 12.2.0.3 版本。按照惯例，这个版本将会支持到 2026 年。

1.2 Oracle 数据库系统的安装

1.2.1 Oracle 19c 的下载

若想要下载 Oracle 数据库软件的最新版本，可以访问 Oracle 公司的官网(https://www.Oracle.com)。

本书以 Windows 10 64 位计算机进行下载演示，其操作步骤和方法如下。

对于企业，可下载 Linux 或者 UNIX 版的 Oracle 19c；但是对于个人用于学习而言，可以选择 Windows 版的 Oracle 19c 即可，可将计算机当作一套数据库服务器。

Windows 版的 Oracle 19c 的下载示意如图 1-1 所示。

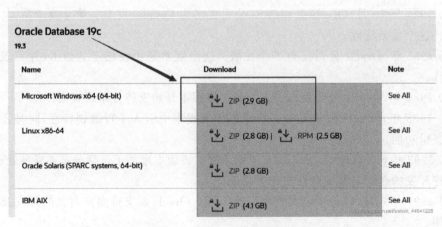

图 1-1　Oracle 19c 下载示意

注意：

（1）由于文件比较大，下载过程中应确保网络稳定，不能中途断网，否则，须重新下载。

（2）在安装前应退出杀毒软件。

（3）单击"下载"按钮，会要求登录账号，如果没有账号，直接注册一个即可。

（4）在压缩包下载完成后，直接解压缩。

（5）在安装前需要把文件包解压，解压后的文件大概需要 6GB 空间，所以要提前确定好解压路径，并确保空间足够大。

1.2.2　Oracle 19c 的安装

本书选择 Oracle 19c 下载的版本为 Microsoft Windows x64，所以下面着重介绍以 Windows 版本作为平台的安装步骤。在 Windows 环境下安装 Oracle 相对比较简单，因为省去了修改各种操作系统参数的步骤。在安装 Oracle 之前，首先应该明确是服务器端的安装还是客户端的安装，下面介绍的是 Oracle 19c 数据库服务器端的安装步骤。

步骤 1　以管理员身份运行启动文件 setup.exe，如图 1-2 所示。在双击后会弹出如图 1-3 所示的"Oracle 19c 安装程序"对话框。

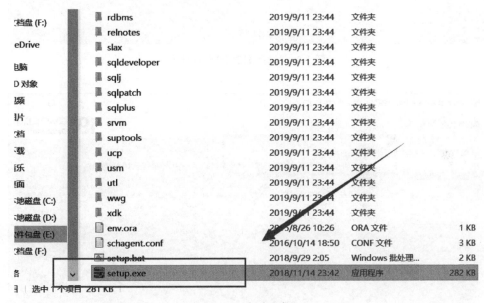

图 1-2　启动文件

步骤 2　在如图 1-3 所示的对话框选择"配置选项"→"创建并配置单实例数据库"选项，单击"下一步"按钮。

步骤 3　在弹出的"Oracle 19c 安装程序-第 2 步（共 15 步）"之"选择系统类"的对话框中，可根据需要选择安装的类型，在个人测试学习环境中，选择"系统类"→"桌面类"选项，可以使得 Oracle 数据库软件允许采用最低的系统配置，再单击"下一步"按钮即可。"系统类"选项卡如图 1-4 所示。

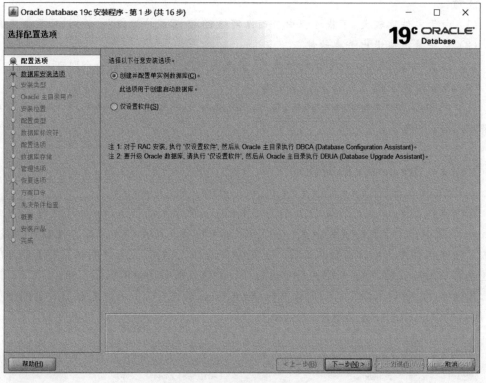

图 1-3 "选择配置选项"选项卡

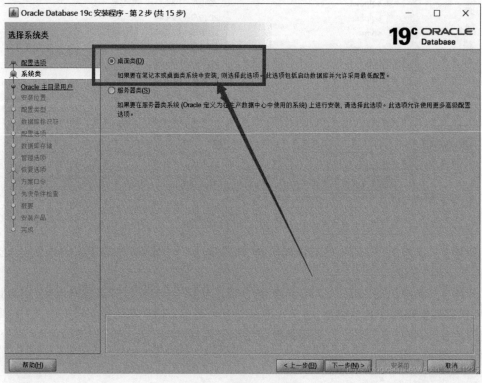

图 1-4 "系统类"选项卡

步骤4　在弹出的"指定 Oracle 主目录用户"对话框中，选择"Oracle 主目录用户"→"创建新 Windows 用户"选项，输入用户名和口令并再次确认后，单击"下一步"按钮，具体如图 1-5 所示。从数据库安全的角度考虑，建议使用非管理员账户来安装配置 Oracle 软件。

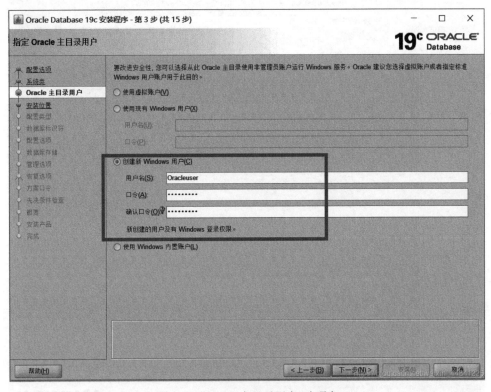

图 1-5　"Oracle 主目录用户"选项卡

步骤5　在弹出的"典型安装配置"对话框中，单击"典型安装"选项卡，填写如下内容，如图 1-6 所示。

（1）在"Oracle 基目录"下拉列表框中选择如 E:\oracle 19c 选项，用于存放所有 Oracle 相关软件的基本目录。

（2）在"数据库文件位置"文本框中输入如 E:\oracle 19c\oradata。

（3）在"字符集"下拉列表框中选择"ZHS 16GBK"选项。

（4）在"全局数据库名"文本框和"口令"文本框中输入相应名称与口令。注意，Oracle 规定了口令的标准，如果密码设计得太过简单，就会出现警告信息，不过可以忽略继续。

（5）取消选择"创建为容器数据库"选项。

进行完以上操作后，单击"下一步"按钮。

步骤6　弹出"先决条件检查"选项卡，主要用于检查计算机是否符合 Oracle 19c 安装的最低配置要求，以及计算机内存空间是否足够，如图 1-7 所示。先决条件满足后即可进入"概要"对话框。

提示：

下面这一步需要关闭杀毒软件，如 360 相关软件等，因为要修改的注册表比较多，安装时间也较长，如果不注意，很容易被禁止修改权限而造成安装出现问题。

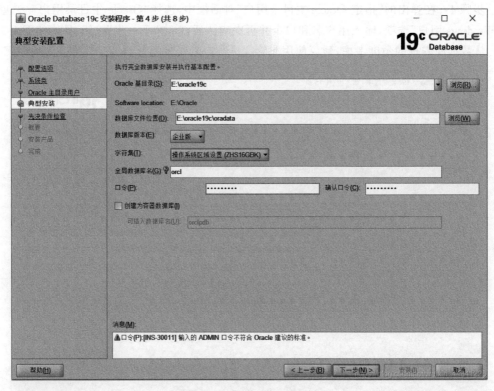

图 1-6 "典型安装"选项卡

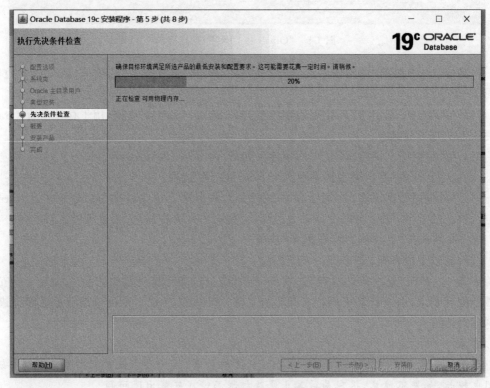

图 1-7 "先决条件检查"选项卡

步骤 7　在"概要"对话框显示安装的概要提示，单击"安装"按钮即可确认安装，如图 1-8 所示。

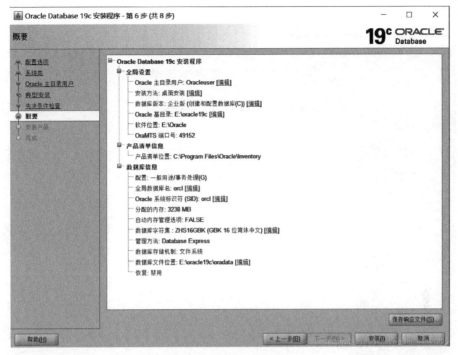

图 1-8　"概要"对话框

步骤 8　"安装产品"对话框表示正式进入安装阶段，在图 1-9 中可以看到 Oracle 19c 的安装进度和状态。

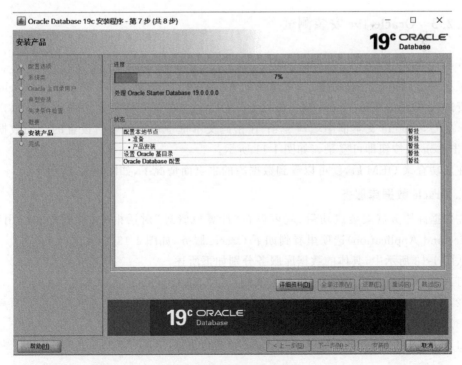

图 1-9　"安装产品"对话框

步骤9　在安装成功后,会弹出如图 1-10 所示的"完成"对话框。单击"关闭"按钮,即可完成安装。

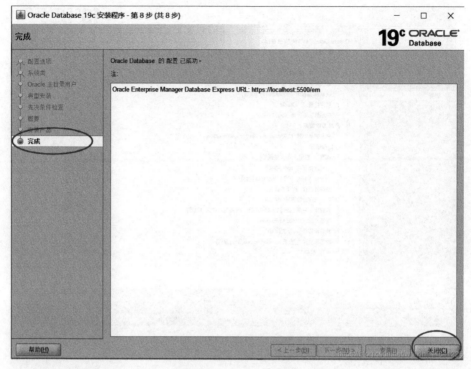

图 1-10　"完成"对话框

1.2.3　Oracle 19c 安装测试

1. Oracle Enterprise Management

根据安装结束后的成功提示,可以访问 https://localhost:5500/em/登录 Oracle 企业资源管理器(Oracle Enterprise Management,OEM)。在打开的页面中需要填写用户名和密码:在 Username 文本框中填写 sys;在密码文本框中填写安装时设置的全局数据库口令,单击 Log in 按钮即可登录。如图 1-11 所示。

在成功登录 OEM 后,就可以看到数据库的主页面情况了,如图 1-12 所示。

2. Oracle 数据库服务

在数据库服务器安装成功后,就可以在"计算机管理"对话框中的"服务和应用程序" (Service and Application)选项里看到如下 Oracle 服务,如图 1-13 所示。

从图 1-13 所示中,具体的数据库服务分别如下所述。

(1) OracleService＜SID＞:Oracle 实例。

(2) Oracle＜HOME_NAME＞TNSListener:Oracle 监听服务。

(3) OracleVssWriter＜SID＞:卷映射复制写入服务。

(4) OracleDBConsole＜SID＞:Oracle 控制台程序。

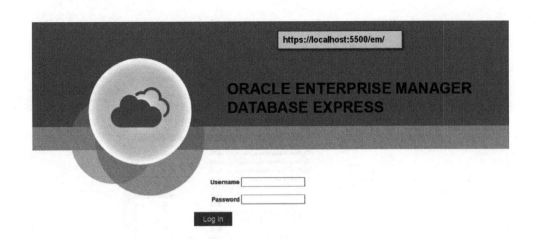

图 1-11　OEM 登录界面

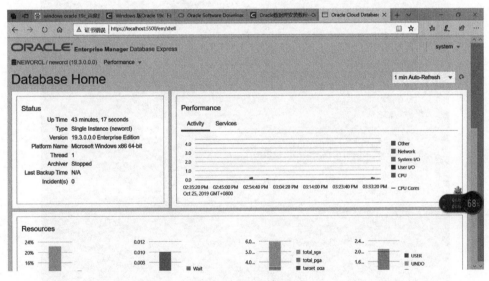

图 1-12　OEM 主页面

（5）OracleJobScheduler＜SID＞：作业调度。

（6）OracleMTSRecoveryService：服务端控制。

注意：在连接 Oracle 之前，必须打开 OracleService＜SID＞和 Oracle＜HOME_NAME＞TNSListener 两个服务。

3. Oracle 的目录结构

在安装数据库结束后，还可以选择"开始"→Oracle-OraDB19Home1 选项，查看 Oracle 应用及服务项，如图 1-14 所示。

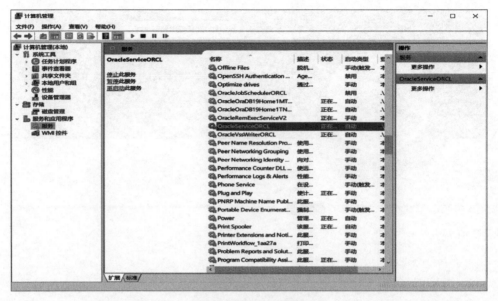

图 1-13　Oracle 服务

图 1-14　Oracle 应用及服务项

4. Oracle 运行测试

在没有安装任何 Oracle 开发和管理相关软件的前提下，可以利用命令窗口进行 Oracle 的运行测试，也可以利用 Oracle 自带的 SQL Plus 软件进行 Oracle 的运行测试。

以管理员身份运行一个"命令提示符"窗口，只须输入用户名和密码即可，如图 1-15 所示。例如，可以使用用户名 system 登录，密码为图 1-6 所示中所填写的密码。若出现如图 1-15 所示的界面，就说明已安装成功了。

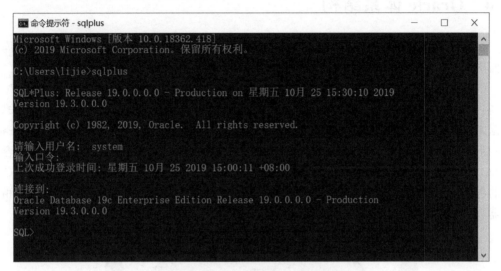

图 1-15　在"命令提示符"窗口中测试 Oracle 登录

1.3　Oracle 数据库服务器

众所周知，Oracle 数据库是一款关系型数据库管理系统。同类的产品还有 DB2、SQL Server 等，通常 Oracle 数据库管理系统被笼统地称为数据库服务器，而数据库服务器又是由实例与数据库两个部分组成。所以一般会说关闭和启动实例，加载和卸载数据库。

Oracle 服务器的组成可以形象化地表示，如图 1-16 所示。

图 1-16　形象化的 Oracle 服务器的组成示意

这里面涉及 3 个主要概念：实例、数据库和数据库服务器。下面先简单对其概念有个初步的认识。

（1）Oracle 实例（Oracle Instance）。一个 Oracle 实例由一系列的后台进程和内存结构组成。换言之，实例是一种访问 Oracle 数据库的机制，由内存和进程结构组成。关于数据库实例会在第 2 章展开讨论。

（2）Oracle 数据库（Oracle Database）。Oracle 数据库是数据的物理存储，包括数据文件、控制文件、联机日志、参数文件等。关于数据库会在第 3 章展开具体讲解。

（3）Oracle 数据库服务器（Oracle Database Server）。Oracle 数据库服务器是一个数据库管理系统，它提供一种开放的、全面的、集成的信息管理方法。它由 Oracle 实例和 Oracle 数据库两部分组成。

1.4　Oracle 体系结构

在了解了数据库的服务器组成后，再来了解数据库的体系结构。观察用户与数据库是如何交互的，交互期间动用了哪些结构？

Oracle 从 11g 过渡到 12c，再到如今的 19c，体系结构也发生了改变。在 Oracle 12c 之前，实例与数据库是一对一或多对一的关系（RAC），即一个实例只能与一个数据库相关联；数据库可以被多个实例所加载。而实例与数据库不可能是一对多的关系。当进入 Oracle 12c 后，实例与数据库可以是一对多的关系。也就是说在 12c 版本后，会在数据库容器（CDB）下创建多个可插拔数据库（PDB），每个 PDB 类似于 11g 里面的实例，然后一个 CDB 下的各个 PDB 是相互隔离的，也就是所谓的多租户体系结构概念。图 1-17 所示为多租户容器数据库（CDBS）体系结构图；图 1-18 为传统数据库（Non-CDBS）体系结构图。

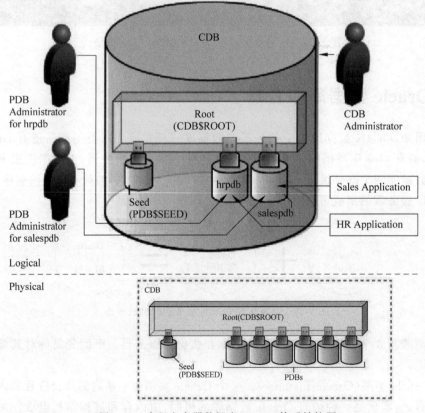

图 1-17　多租户容器数据库（CDBS）体系结构图

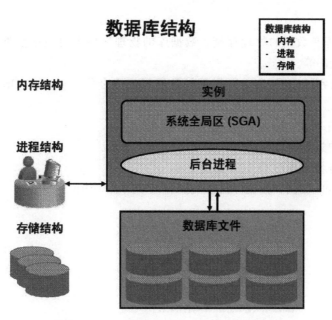

图 1-18　传统数据库（Non-CDBS）体系结构图

本书以传统数据库为讲授要点，在其他相关章节展开知识讨论。从 Oracle 数据库服务器的体系结构图中，Oracle 的体系结构可以从 4 个组成部分进行讲解：用户进程、服务器进程、Oracle 实例与数据库。

1．用户进程

Oracle 的进程分为用户进程、服务器进程和后台进程。其中，后台进程包含在 Oracle 实例里；用户进程是在客户机内存上运行的程序，在用户请求连接 Oracle 数据库时建立。它可以是多种多样的，例如常用的用户进程有：SQL Plus、PL/SQL Develop、OEM、Java 语言编写的应用程序和 Pro C 编写的程序等。这些程序在运行状态下就称为用户进程，它们都通过标准的数据库访问 API 接口，并与 Oracle 服务器进行交互。

2．服务器进程

与用户进程对应的进程就是服务器进程，服务器进程就是接收客户端发来的请求（如查询、修改等），并监护着完成这类请求的服务端程序。服务器进程直接同 Oracle 服务器交互，执行用户请求并返回结果，所以服务器进程是 Oracle 服务器对外交互的接口，相当于 Oracle 服务器的"外交部"。服务器进程主要用来分析和执行 SQL 语句，当所需的数据不在系统全局区（System Global Area，SGA）内存中时，还需要完成从磁盘数据文件复制到系统全局区（SGA）的共享数据缓冲区等工作。

3．Oracle 实例

若要服务器进程与 Oracle 数据库服务器实现交互，必须通过访问数据库文件的唯一通道——Oracle 实例。Oracle 实例就是 Oracle 利用后台内存结构和后台进程管理 Oracle 数据库并提供服务的。它是一组 Oracle 的后台进程以及在服务器中分配的共享存储区域。

4. 数据库

Oracle 数据库是数据的物理存储。数据库的物理文件主要包括数据文件、控制文件、联机日志、参数文件等。

用户进程、服务器进程、Oracle 实例与数据库通力合作即可完成 Oracle 的工作流程。其操作步骤如下：

步骤1　用户与用户进程交互。
步骤2　用户进程与服务器进程交互。
步骤3　服务器进程与 Oracle 实例交互。
步骤4　Oracle 实例和数据库进程交互。

第2章

Oracle实例

PPT 视频讲解

若要了解数据库实例,必须学习数据库的内存结构和进程结构。本章部分实践无须安装数据库开发和管理工具,直接使用命令行即可。

2.1 数据库实例

Oracle 实例是指操作系统中一系列的进程和进程所分配的内存块。通俗地说,Oracle 实例是访问数据库文件的通道,其是一个动态运行的概念。本章从内存和进程两个方面介绍 Oracle 实例的基本概念和原理。

2.1.1 内存结构

内存结构是 Oracle 数据库体系中最重要的一个部分,也是影响数据库性能的第一要素。内存结构主要包括系统全局区和程序全局区。

1. 系统全局区

系统全局区(System Global Area,SGA)是所有用户都可以访问的 Oracle 实例的共享内存区域,是 Oracle 系统为 Oracle 实例分配的一组共享缓冲存储区,用于存放数据库数据和控制信息,以实现对数据库数据的管理和操作。当启动实例时,Oracle 为 SGA 分配内存;当中止实例时,释放 SGA 占用的内存。

SGA 由数据库高速缓冲区、共享池、重做日志缓冲区与大池组成。

2. 程序全局区

程序全局区(Program Global Area,PGA)是一类没有共享的内存,它专用于特定的服务器进程,并且只能由这个进程访问。PGA 是为单独的服务进程存储私有数据的内存区域。

3. PGA 和 SGA 的区别

两者类似,都是 Oracle 数据库系统为会话在服务器内存中分配的区域。不过两者的作用不同,共享程度也不同。SGA 对系统内的所有进程都是共享的;而 PGA 则主要为某个用户进程服务的。

2.1.2 进程结构

进程是操作系统中的一种机制,它可执行一系列操作步骤。操作系统会使用多个进程执行 Oracle 的不同部分,并且对于每一个连接的用户都有一个进程。

进程可分为用户进程和 Oracle 进程。其中,Oracle 进程又可以分为服务器进程和后台进程。

2.2 数据库内存结构

数据库内存结构是 Oracle 存放常用信息和所有运行在该机器上的 Oracle 程序的内存区域。

图 2-1 所示为数据库内存结构图。

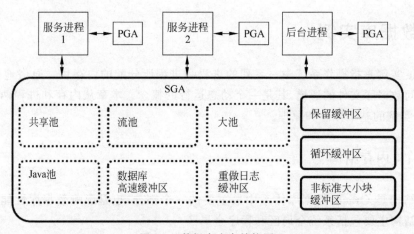

图 2-1 数据库内存结构图

2.2.1 SGA

SGA 是一组为系统分配的共享内存结构,可以包含一个数据库实例的数据或控制信息。如果多个用户连接到同一个数据库实例,在 Oracle 实例的 SGA 中,数据可以被多个用户共享,当数据库实例启动时,SGA 的内存被自动分配;当数据库实例关闭时,SGA 内存被回收。SGA 是占用内存最大的一个区域,同时也是影响数据库性能的重要因素。

1. SGA 的组成

选择"开始"→"运行"选项,输入 cmd 命令,以管理员的身份打开"命令提示符"窗口。

输入 sqlplus，便可以登录 SQL Plus，然后输入登录命令即可连接到 Oracle 服务器上，连接后可以进行内存结构的查询，显示效果如下：

```
C:\Users\lijie>sqlplus

SQL*Plus: Release 19.0.0.0.0 - Production on 星期二 3月 17 20:17:03 2020
Version 19.3.0.0.0

Copyright (c) 1982, 2019, Oracle. All rights reserved.
请输入用户名: sys as sysdba
输入口令:
连接到:
Oracle Database 19c Enterprise Edition Release 19.0.0.0.0 - Production
Version 19.3.0.0.0
SQL> show sga

Total System Global Area    2550136752 bytes
Fixed Size                     9031600 bytes
Variable Size                603979776 bytes
Database Buffers            1929379840 bytes
Redo Buffers                   7745536 bytes
```

SGA 的组成由固定区域、可变区域、数据库高速缓冲区缓存和重做日志缓冲区缓存组成。它们分别如下所述：

(1) 固定区域(Fixed Size)：引导区域，存储 SGA 中各个组件的信息，大小不能修改。

(2) 可变区域(Variable Size)：包括共享池、大池、Java 池。

(3) 数据库高速缓冲区缓存(Database Buffer Cache)：大小由参数 db_cache_size 指定。

(4) 重做日志缓冲区缓存(Redo Log Buffer Cache)：大小通常大于参数 log_buffer 的设置。

SGA 的组成如图 2-2 所示。

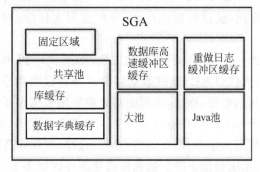

图 2-2　SGA 的组成

若要从动态性能字典中查询 SGA 更加详细的信息,则需输入如下查询命令。

```
SQL>select * from v$sgainfo;

NAME                                    BYTES  RES  CON_ID
--------------------------------   ----------  ---  ------
Fixed SGA Size                        9031600  No        0
Redo Buffers                          7745536  No        0
Buffer Cache Size                  1929379840  Yes       0
In-Memory Area Size                         0  No        0
Shared Pool Size                    520093696  Yes       0
Large Pool Size                      33554432  Yes       0
Java Pool Size                       16777216  Yes       0
Streams Pool Size                    33554432  Yes       0
Shared I/O Pool Size                134217728  Yes       0
Data Transfer Cache Size                    0  Yes       0
Granule Size                         16777216  No        0

NAME                                    BYTES  RES  CON_ID
--------------------------------   ----------  ---  ------
Maximum SGA Size                   2550136752  No        0
Startup overhead in Shared Pool      29526821  No        0
Free SGA Memory Available                   0  0

已选择 14 行。
```

2. 数据库高速缓冲区

数据库高速缓冲区(Database Buffer Cache)为 SGA 的主要成员,用来存放读取自数据文件的数据块副本,或是曾经处理过的数据,为所有与 Oracle 实例相链接的用户进程所共享。

如果用户发出如下指令:select * from dual,那么 Oracle 是怎样提取数据库中的数据呢? 服务器进程将首先在数据库高速缓冲区中搜索所需数据,如果找到了,就直接使用而不进行磁盘操作;如果没找到,就进行磁盘操作,并把数据文件中的数据读入数据库高速缓冲区。

数据库高速缓冲区的大小可以由服务器文件 spfile.ora 中的 db_cache_size 参数设定。通常,采用最近最少使用算法(LRU)管理数据库高速缓冲区可用空间。如果 SGA 的大小不足以容纳所有最常使用的数据,那么,不同的对象将争用数据库高速缓冲区中的空间。当多个应用程序共享同一个 SGA 时,很有可能发生这种情况。此时,每个应用的最近使用段都将与其他应用的最近使用段争夺 SGA 中的空间。这样,数据库高速缓冲区的数据请求将出现较低的命中率,导致系统性能下降。数据库高速缓冲区如图 2-3 所示。

通过以下命令可以在 SQL Plus 中查询或管理数据库缓存。

(1) 显示数据库高速缓冲区的大小:SQL>show parameter db_cache_size。

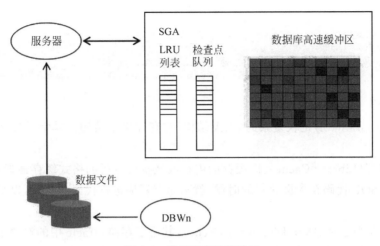

图 2-3 数据库高速缓冲区

(2) 显示保留缓冲区的大小：SQL＞show parameter db_keep_cache_size。
(3) 显示循环缓冲区的大小：SQL＞show parameter db_recycle_cache_size。
(4) 修改数据库高速缓冲区的大小：SQL＞alter system set db_cache_size＝500m。
(5) 清空数据库高速缓冲区：SQL＞alter system flush buffer_cache。

3. 重做日志缓冲区

联机重做日志文件用于记录数据库的更改，对数据库进行修改的事务在记录到重做日志之前须放到重做日志缓冲区(Redo Log Buffer)。SGA 中也分配一片内存区用于缓存重做日志，这片内存区称为重做日志缓冲区。事务日志先写入重做日志缓冲区，待一定的时机由 LGWR 将重做日志缓存中的信息写入联机日志文件。

重做日志缓冲区如图 2-4 所示。

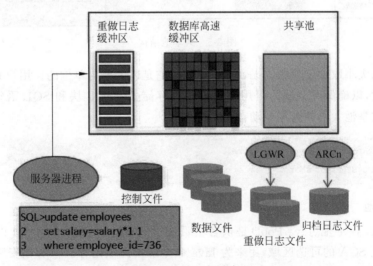

图 2-4 重做日志缓冲区

重做日志缓冲区的大小由 log_buffer 确定。若要显示以及动态调整重做日志缓冲区大小，指令如下。

```
SQL>show parameter log_buffer
SQL>alter system set log_buffers = 65536 scope = spfile
```

4. 共享池

共享池(Share Pool)是较复杂的 SGA 结构，它有许多子结构。其常见的 3 个共享池组件如下所述。

(1) 库缓存(Library Cache)：库缓存的内存区域按已分析的格式缓存最近执行的代码，这样，同样的 SQL 代码在多次执行的时候，就不用重复地进行代码分析，可以很大程度上提高系统性能。

(2) 数据字典缓存(Data Dictionary Cache)：其用于存储 Oracle 中的对象定义(包括表、视图、同义词和索引等数据库对象)。这样，在分析 SQL 代码的时候，就不用频繁地从磁盘上读取数据字典中的数据。

(3) PL/SQL 区：其用于缓存存储过程、函数、触发器等数据库对象。这些对象都存储在数据字典中，通过将其缓存到内存中，可以在重复调用的时候提高性能。

共享池结构如图 2-5 所示。

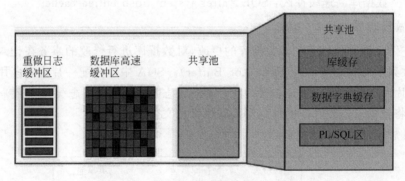

图 2-5　共享池结构

共享池的大小取决于参数 shared_pool_size，它是以字节为单位的。用户必须将这个值设置得足够大，以确保有足够的可用空间装载和存储 PL/SQL 块和 SQL 语句。可利用如下命令进行共享池大小的查询与动态调整。

```
SQL>show parameter shared_pool_size;
SQL>alter system set shared_pool_size = 500m;
```

5. Java 池

SGA 内的 Java 池(Java Pool)是供各会话内运行的 Java 代码及 JVM 内的数据使用的。Java 池是 SGA 的可选区域，用来为 Java 命令解析提供内存。只有在安装和使用 Java 时才需要 Java 池。Java 池的大小以字节为单位，由 java_pool_size 参数设置，java_pool_size 参数设置默认为 10MB。

6. 大池

大池(Large Pool)是一个可选内存区。如果使用线程服务器选项或者频繁地执行备份与恢复操作,则只要创建一个大池就可以更有效地管理这些操作了。大池必须显式配置。大池的内存不是来自共享池,而是直接来自 SGA,这就增大了 Oracle 服务器在 Oracle 实例启动时需要的共享内存量。大池的大小由 large_pool_size 参数指定。

2.2.2 PGA

程序全局区(PGA)是一个非共享的内存区域,其中包含专门供服务器和后台进程使用的数据和控制信息。在专用服务器环境中,为每个服务器和启动的后台进程创建一个 PGA。每个 PGA 都由堆栈空间、哈希区、位图合并区和用户全局区(UGA)组成。当终止与之关联的服务器或后台进程时,将释放 PGA。PGA 的结构如图 2-6 所示。

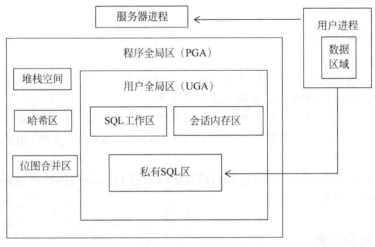

图 2-6　PGA 的结构

在共享服务器环境中,多个客户端用户共享服务器进程。UGA 被移入大型池,而 PGA 仅具有堆栈空间、哈希区和位图合并区。

在专用服务器会话中,PGA 由以下组件组成。

(1) 堆栈空间:堆栈空间能够分配用于保存会话变量和数组的内存。

(2) 哈希区:此区域用于执行表的哈希连接。

(3) 位图合并区:此区域用于合并从多个位图索引的扫描中检索到的数据。

(4) 用户全局区。

① SQL 工作区:此区域供排序数据的函数使用,如 order by 和 group by。

② 会话内存区:此用户会话数据存储区将分配给会话变量,如登录信息和数据库会话所需的信息。

③ 私有 SQL 区:此区域保存与之相关的已解析的 SQL 语句的信息以及其他特定会话的信息以供处理。

在专有服务器(Dedicated Server)模式下,Oracle 会为每个会话启动一个 Oracle 进程;

而在共享服务器(Shared Server)模式下,由多个会话共享同一个 Oracle 服务进程。

专用服务器模式与共享服务器模式下的 PGA,如图 2-7 所示。

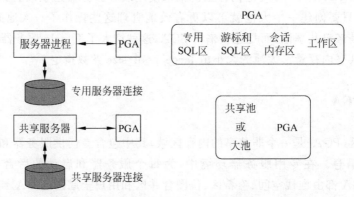

图 2-7　专用服务器模式与共享服务器模式下的 PGA

2.3　数据库进程

数据库进程的主要作用是维护数据库的稳定,相当于一个企业中的管理者,具有负责全局统筹的作用。Oracle 数据库进程共分为以下 3 类。

(1) User Process(用户进程):在一个数据库用户请求与 Oracle 服务器连接时启动。

(2) Server Process(服务器进程):与 Oracle Instance(Oracle 实例)相连,当用户创建会话时启动。

(3) Background Process(后台进程):后台进程伴随 Oracle 实例的启动而启动,它们并不直接给用户提供服务。

2.3.1　用户进程

用户进程指的是用户在运行程序或者 Oracle 工具时,需要通过建立用户进程和 Oracle 实例进行通信。常说的 Connection(连接)就是用户进程和 Oracle 实例间建立的一个通信通道。Oracle 的连接允许一个用户同时多次连接到同一个数据库实例。常说的会话(Session)是用户在和 Oracle 服务器连接成功,并通过了 Oracle 的身份验证后,用户会与 Oracle 服务器之间建立一个会话。同时,同一个用户可以并发地与数据库建立多个会话。用户进程示例如图 2-8 所示。

要了解用户进程相关信息,可以使用动态性能视图(v$session)。它主要提供数据库连接的信息,主要是客户端的信息,观察该动态视图结构,了解其主要的属性列。

(1) machine:在哪台机器上。

(2) terminal:使用什么终端。

(3) osuser:操作系统用户是谁。

(4) program:通过什么客户端程序,例如 SQL Plus。

(5) process:操作系统分配给 SQL Plus 的进程号。

(6) taddr:是否执行事务处理中。

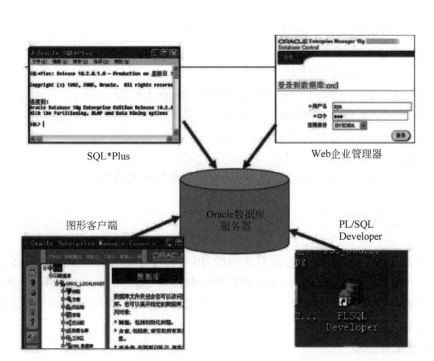

图 2-8 用户进程示例

当然,v$session 也有一些是服务器端的信息:

(1) paddr:即 v$process 中的服务器进程的 addr。
(2) server:服务器是专有服务器还是共享服务器。

2.3.2 服务器进程

服务器进程主要用来分析和执行 SQL 语句,当所需的数据不在 SGA 内存中时,就要执行从磁盘数据文件复制到 SGA 的共享数据缓冲区等工作。

用户进程与服务进程的关系图,如图 2-9 所示。

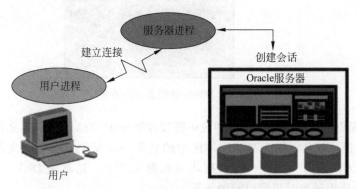

图 2-9 用户进程与服务进程的关系图

Oracle 数据库提供了相应的视图 v$process,以用来监视整个数据库的进程情况。该动态视图中的常用属性列如下所述。具体的使用将在本章后续的案例中讲解。

(1) addr：可以与 v$session 的 paddr 字段关联。
(2) pid：Oracle 进程标志。
(3) spid：操作系统进程标志。
(4) pname：进程的名称（从中可以看到前台和后台进程）。
(5) userame：运行此进程的操作系统的用户名称。
(6) terminal：终端名称类似于 v$session。
(7) program：程序的名称。
(8) background：1 代表后台进程；否则代表正常服务器进程。

2.3.3 后台进程

后台进程是在 Oracle 实例启动时建立的，用于优化性能和协调多用户连接通信的工作。常用的 Oracle 后台进程，可以通过数据字典中的 v$bgprocess 和 v$process 查询。

后台进程中有 5 个重要也是必须执行的后台进程，当数据库启动后，这 5 个进程自动于后台开始执行，下面详细介绍。

1. 数据库写进程

数据库的典型操作就是大规模的 I/O，若要提高 Oracle 系统的效率，就要减少执行 I/O 的次数。这也是 Oracle 引入数据库写（Database Writer，DBWn）进程的主要原因之一。

DBWn 进程的工作原理如图 2-10 所示。

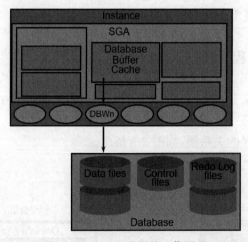

图 2-10 DBWn 进程的工作原理

DBWn 进程负责将数据库高速缓冲区中脏缓冲区中的数据写到数据文件上，为了提高效率，DBWn 进程并不是数据库高速缓冲区中的数据一有变化就写数据文件，而是积累了足够多的数据，才依次写一大批内存数据块到数据文件上。这就是 DBWn 进程的触发时机，具体 DBWn 进程触发时机可以归纳如下。

(1) 在发生检查点（CKPT）进程时，触发 DBWn 进程。
(2) 如果脏数据块的总数超过一定限度，就触发 DBWn 进程。
(3) 如果服务进程在设定的时间内没有找到空闲块，就触发 DBWn 进程。

（4）每 3s 自动唤醒一次 DBWn 进程。

Oracle 通常使用 db_writer_process 定义 DBWn 进程的数量，如果在 Oracle 实例启动时没有说明 DBWn 进程的个数，Oracle 将根据 CPU 的数量决定参数的设置。执行下面的查询，可以显示进程的个数。

```
SQL> show parameter db_writer_process;
NAME                                 TYPE        VALUE
------------------------------------ ----------- ----------
db_writer_processes                  integer     1
```

2. 检查点进程

Oracle 系统为了提高系统的效率和维护数据库的一致性，引入了一个称为检查点（Check Point Process，CKPT）进程。CKPT 进程负责发起检查点信号。检查点可以强制 DBWn 进程写入缓冲区并更新控制文件。

CKPT 进程的职责可以细化如下所述。

（1）用检查点信息修改数据文件头，改写新的 SCN。

（2）用检查点信息修改控制文件，改写最新的 SCN。

（3）在检查点时调用 DBWn 进程，通知其将缓存中的脏数据写入磁盘。

CKPT 进程的工作原理如图 2-11 所示。

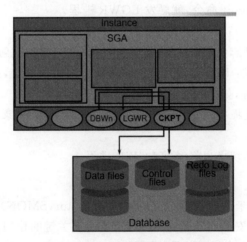

图 2-11　CKPT 进程的工作原理

CKPT 进程的触发时机可以归纳如下所述。

（1）用户进程任意的事务提交、任何的 DDL 与 DCL 都将触发 CKPT 进程。

（2）日志的切换将触发 CKPT 进程。

若要手动设置检查点，则执行如下代码。

```
SQL> alter system checkpoint;
系统已更改.
```

3. 日志写进程

日志写(Log Write,LGWR)进程也是一种后台进程,主要负责将日志缓冲内容写到磁盘的在线重做日志文件或重做日志文件组中。

LGWR 进程的工作原理如图 2-12 所示。

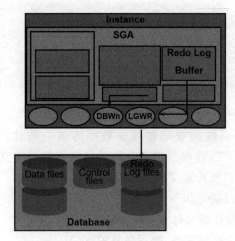

图 2-12　LGWR 进程的工作原理

LGWR 进程的触发条件可以归纳如下。

(1) 若用户发出 commit 命令,就触发 LGWR 进程。

(2) 每 3s 定时唤醒 LGWR 进程。

(3) 若日志缓冲区日志量超过 1/3,或日志数量超过 1MB,就触发 LGWR 进程。

(4) DBWn 进程触发:DBWn 进程试图将脏数据块写入磁盘,并检测他的相关 redo 记录是否写入联机日志文件,如果没有,就通知 LGWR 进程。

注意:在 Oracle 中是提前写机制(Write Ahead),即 redo 记录须先于数据记录被写入磁盘。

(5) 联机日志文件的切换也将触发 LGWR 进程。

4. 系统监控进程

对于 Oracle 数据库来说,系统监控(System Monitor,SMON)进程负责的内容并不是很多,但是对于数据库的安全与数据库的性能却有着很关键的作用。

SMON 进程的主要职责可以归纳如下所述。

(1) 负责 Oracle 实例恢复。前滚(Roll Forward)恢复到 Oracle 实例关闭的状态,使用最后一次检查点后的日志进程重做。这时包括提交和未提交的事务。打开数据库,进行回滚(Roll Back),即回滚未提交的事务。

(2) 负责清理临时段,以释放空间。

SMON 进程的工作原理如图 2-13 所示。

SMON 进程的触发条件有定期被唤醒和被其他事务主动唤醒两个。

5. 进程监视进程

进程监视(Process Monitor,PMON)进程主要用来监视服务器进程。当某个进程崩溃

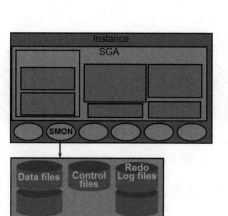

图 2-13　SMON 进程的工作原理

时,PMON 进程将负责它的清理工作。主要包括回滚用户当前的事物、释放用户所占用的锁、释放用户所有的其他资源等。

PMON 进程的工作原理如图 2-14 所示。

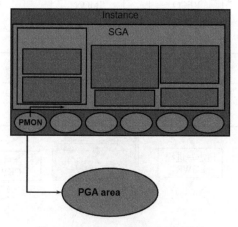

图 2-14　PMON 进程的工作原理

PMON 进程的触发条件有定时被唤醒和被其他进程主动唤醒两种。

6. 其他进程

以上 5 个后台进程都是必须的,也就是说它们中的任何一个停止,Oracle 实例将会自动关闭,在可选后台进程中还有其他进程,其中最主要的就是归档日志进程。

(1) 归档日志(Archive Log,ARCn)进程。

ARCn 进程的职责是发生日志切换时把写满的联机日志文件复制到归档目录中。写日志写到需要覆盖重写的时候,触发 ARCn 进程去转移日志文件,进行复制后形成归档日志文件,以免日志丢失。ARCn 进程会在日志切换时被 LGWR 进程唤醒。

ARCn 进程的工作原理如图 2-15 所示。

(2) RECO:恢复进程,维护分布式数据库环境下的一致性。

(3) LCKn:锁进程,用于多个 Oracle 实例之间的封锁。

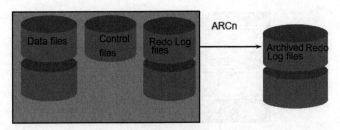

图 2-15　ARCn 进程的工作原理

（4）Dnnn：调度进程，用于多线程服务器模式下。

2.4　进程与内存的交互情况

进程与内存的交互情况可用其协作关系表示。进程与内存结构、数据文件间的协作关系如图 2-16 所示。

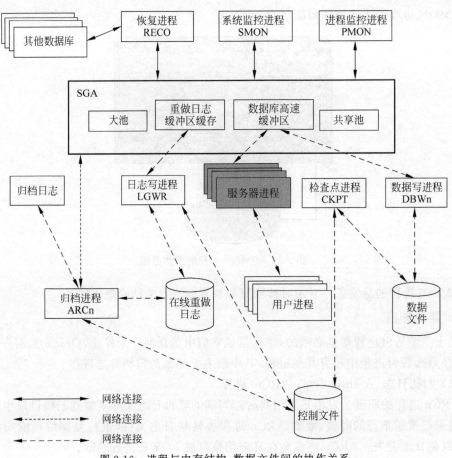

图 2-16　进程与内存结构、数据文件间的协作关系

2.5 项目案例

2.5.1 用户进程实例

视频讲解

下面通过实验过程观察用户进程的相关操作。

步骤1 查看任务管理器中的初始系统进程,如图 2-17 所示。观察当前操作系统进程。发现当前使用命令行运行 SQL Plus。

名称	PID	状态	用户名	CPU	内存(活动的...	UAC 虚拟化
sqlplus.exe	12916	正在运行	lijie	00	2,028 K	不允许
sqlservr.exe	4880	正在运行	NETWORK...	00	5,808 K	不允许

图 2-17 任务管理器中的初始系统进程

步骤2 打开 SQL Plus 应用程序,使用用户 hr 登录。
在第一次使用用户 hr 登录时,必须在管理员用户下对用户 hr 进行解锁和修改密码。

```
SQL> alter user hr identified by hr account unlock;
用户已更改.
SQL> conn hr
输入口令:
已连接.
```

步骤3 再次查看任务管理器中的系统进程,如图 2-18 所示,观察操作系统进程的变化。

名称	PID	状态	用户名	CPU	内存(活动的...	UAC 虚拟化
spoolsv.exe	3796	正在运行	SYSTEM	00	516 K	不允许
sqlplus.exe	33772	正在运行	lijie	00	4,892 K	不允许
sqlplus.exe	12916	正在运行	lijie	00	2,028 K	不允许
sqlservr.exe	4880	正在运行	NETWORK...	00	5,788 K	不允许

图 2-18 任务管理器中的系统进程

步骤4 使用 v$session 会话进程也可以查看用户进程相关链接信息。
v$session 主要提供的是一个数据库连接的信息,主要是客户端的信息。代码如下。

```
SQL> col sid format 9999
SQL> col process format a30
SQL> col machine format a30
SQL> col process format a30
SQL> col username format a20
SQL> select sid,username,process,machine,program from v$session;

  SID USERNAME   PROCESS     MACHINE              PROGRAM
----- --------   ---------   ------------------   -------------------
    1            37744       LAPTOP-8BM7LBRI      ORACLE.EXE (DBRM)
```

```
    2                25080           LAPTOP-8BM7LBRI         ORACLE.EXE (LGWR)
    4                40396           LAPTOP-8BM7LBRI         ORACLE.EXE (TMON)
    6                32056           LAPTOP-8BM7LBRI         ORACLE.EXE (TT00)
   10                17996           LAPTOP-8BM7LBRI         ORACLE.EXE (W00E)
   11                16120           LAPTOP-8BM7LBRI         ORACLE.EXE (W001)
   12                36452           LAPTOP-8BM7LBRI         ORACLE.EXE (W006)
   13 SYS            12916:28664     WORKGROUP\LAPTOP-8BM7LBRI    sqlplus.exe
  255 HR             33772:13776     WORKGROUP\LAPTOP-8BMLBRI     sqlplus.exe
......
已选择 64 行.
```

加上用户名这个条件即可清晰地查看到进程的详细信息,具体代码如下:

```
SQL> select process,machine,program,saddr,paddr from v$session where username = 'HR';
PROCESS         MACHINE                     PROGRAM         SADDRPADDR
--------------- --------------------------- --------------- ------------------
33772:13776     WORKGROUP\LAPTOP-8BM7LBRI   sqlplus.exe     00007FFEE79E8828
00007FFEE743C748
```

步骤 5 在 v$session 动态视图中,taddr 表示是否执行事务处理中。在执行下面的代码时着重观察 taddr 的值,发现其值为空。

```
SQL> select sid,taddr,process,machine,terminal,program from v$session where username = 'HR';

SID  TADDR      PROCESS          MACHINE                      TERMINAL            PROGRAM
---- ---------- ---------------- ---------------------------- ------------------- -----------
255             33772:13776      WORKGROUP\LAPTOP-8BM7LBRI    LAPTOP-8BM7LBRI     sqlplus.exe
```

步骤 6 在用户 hr 下执行 DML 语句。

```
SQL> update employees set salary = salary + 100;

已更新 107 行.
```

步骤 7 修改数据后,再次观察 taddr 的值。

```
SQL> select sid,taddr,process,machine,terminal,program from v$session where username = 'HR';
SID TADDR              PROCESS          MACHINE
---- ----------------- ---------------- ---------------------------
TERMINAL            PROGRAM
------------------- -----------
255 00007FFEDF189D70  33772:13776      WORKGROUP\LAPTOP-8BM7LBRI
LAPTOP-8BM7LBRI     sqlplus.exe
```

执行更新操作后,再次查询 v$session 时发现 tdaar 值非空,这代表这个进程正在执行事务处理。

步骤 8 在用户 hr 下回滚刚刚执行的 DML 操作。

```
SQL> rollback
回滚已完成.
```

步骤 9 再次观察 taddr 的值,若发现 taddr 的值又再次为空,则代表该进程无事务处理。

```
SQL> select sid,taddr,process,machine,terminal,program from v$session where username='HR';

   SID TADDR    PROCESS      MACHINE                 TERMINAL           PROGRAM
------ -------- ------------ ----------------------- ------------------ -----------
   255          33772:13776  WORKGROUP\LAPTOP-8BM7LBRI LAPTOP-8BM7LBRI  sqlplus.exe
```

2.5.2 服务器进程实例

视频讲解

下面通过实验感受一下后台进程有哪些？它们又是如何与 Oracle 数据库的不同部分进行交互的？

步骤 1 利用 v$process 的 background 属性查找 Oracle 启动的所有后台进程。

```
SQL> select pid,pname,program from v$process where background='1';

   PID PNAME PROGRAM
------ ----- -----------------
     2 PMON  ORACLE.EXE (PMON)
     3 CLMN  ORACLE.EXE (CLMN)
     4 PSP0  ORACLE.EXE (PSP0)
     5 VKTM  ORACLE.EXE (VKTM)
 ......
已选择 62 行.
```

步骤 2 利用 v$bgprocess 的 paddr 属性查看所有后台进程。

```
SQL> col description for a30
SQL> select paddr,name,description from v$bgprocess where paddr<>'00';

PADDR            NAME  DESCRIPTION
---------------- ----- ------------------------------
00007FFEE73D02C8 PMON  process cleanup
00007FFEE73D1870 CLMN  process cleanup
00007FFEE73D2E18 PSP0  process spawner 0
00007FFEE73D43C0 VKTM  Virtual Keeper of TiMe process
00007FFEE73D5968 GEN0  generic0
00007FFEE73D6F10 MMAN  Memory Manager
```

```
00007FFEE73D84B8    DBRM    DataBase Resource Manager
00007FFEE73D9A60    GEN1    generic1
00007FFEE73DB008    DIAG    diagnosibility process
00007FFEE73DC5B0    VKRM    Virtual sKeduler for Resource
                            Manager
......
已选择 81 行.
```

步骤 3 利用 v$bgprocess 与 v$process 查看所有的后台进程情况。

```
SQL> select p.spid,b.name,b.description from v$process p, v$bgprocess b
where p.addr = b.paddr;

SPID                    NAME    DESCRIPTION
--------------------    ----    --------------------
33728                   PMON    process cleanup
452                     CLMN    process cleanup
19768                   PSP0    process spawner 0
14440                   VKTM    Virtual Keeper of TiMe process
84                      GEN0    generic0
1304                    MMAN    Memory Manager
37744                   DBRM    DataBase Resource Manager
31840                   GEN1    generic1
26916                   DIAG    diagnosibility process
23236                   VKRM    Virtual sKeduler for Resource
                                Manager
.......
已选择 81 行.
```

2.5.3 SGA 实例

通过反复执行同一个查询,验证数据库高速缓冲区和共享池的作用。具体操作步骤如下。

步骤 1 设定 SQL Plus 显示时间,方便用户查看该查询执行的所用时间。

```
SQL> set timing on
```

步骤 2 以表 dba_objects 的聚组函数为例进行第一次查询,观察执行所用时间。

```
SQL> select count(*) from dba_objects;

COUNT(*)
```

```
----------
72728
```

已用时间：00：00：00.22

步骤 3　以表 dba_objects 的聚组函数为例进行第二次查询，观察执行所用时间。

```
SQL> select count( * ) from dba_objects;

COUNT( * )
----------
72728
```

已用时间：00：00：00.05

步骤 4　清空共享缓存。

```
SQL> alter system flush shared_pool;

系统已更改.
```

已用时间：00：00：00.40

步骤 5　清空数据库块缓存。

```
SQL> alter system flush buffer_cache;

系统已更改.
```

已用时间：00：00：00.06

步骤 6　在清空共享内存和数据块缓存下，以表 dba_objects 的聚组函数为例进行第三次查询，观察执行所用时间。

```
SQL> select count( * ) from dba_objects;

COUNT( * )
------------------
72728
```

已用时间：00：00：00.52

第3章

Oracle数据库

PPT 视频讲解

3.1　Oracle 数据库概述

在本书第 1 章中曾指出，Oracle 服务器是由静态的数据库和动态的 Oracle 实例组成的。本章将介绍 Oracle 数据库。

数据库是数据存储的容器，用来收集、存储数据和返回信息。Oracle 数据库可以从逻辑存储结构和物理存储结构两个角度来划分层次。

物理存储结构主要用于描述在 Oracle 数据库外部数据的存储，即在操作系统层面中如何组织和管理数据，与具体的操作系统有关；逻辑存储结构主要描述 Oracle 数据库内部数据的组织和管理方式，即在数据库管理系统的层面中如何组织和管理数据，与操作系统没有关系。物理存储结构具体表现为一系列的操作系统文件，是可见的；逻辑存储结构是物理存储结构的抽象体现，是不可见的，可以通过查询数据库数据字典来了解逻辑存储结构。

Oracle 数据库的物理存储结构与逻辑存储结构既相互独立又相互联系，它们之间的关系如图 3-1 所示。

从图 3-1 中可以看出，Oracle 数据库的物理存储结构和逻辑存储结构的基本关系可以归纳如下所述。

(1) 一个数据库在物理上包含多个数据文件，在逻辑上包含多个表空间。

(2) 一个表空间包含一个或多个数据文件，一个数据文件只能从属于某个表空间。

(3) 一个区只能从属于一个数据文件，而一个数据文件可包括一个或多个区。

(4) Oracle 数据库的块由一个或多个操作系统块组成。

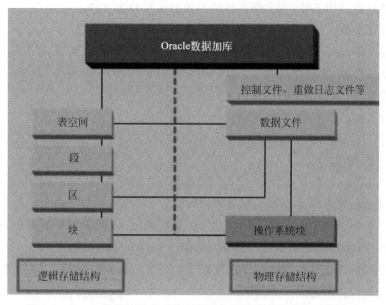

图 3-1 Oracle 数据库的物理存储结构与逻辑存储结构的关系示意

3.2 Oracle 数据库物理存储

Oracle 数据库物理存储是存储在磁盘中的操作系统文件，它的组成如图 3-2 所示。

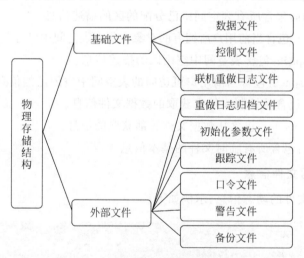

图 3-2 Oracle 数据库物理存储的组成

其中最主要的 3 种类型文件分别为数据文件、控制文件和联机重做日志文件，这三大核心文件对 Oracle 数据库的正常启动是缺一不可的。具体的物理存储结构组成文件如下。

(1) 数据文件：用于存储数据库中的所有数据。

(2) 控制文件：用于记录和描述数据库的物理存储结构信息。

(3) 联机重做日志文件：用于记录外部程序（用户）对数据库的改变操作。

(4) 重做日志归档文件:用于保存已经写满的重做日志文件。
(5) 初始化参数文件:用于设置数据库启动时的参数初始值。
(6) 跟踪文件:用于记录用户进程、数据库后台进程等的运行情况。
(7) 口令文件:用于保存具有 sysdba、sysoper 权限的用户名和用户口令。
(8) 警告文件:用于记录数据库的重要活动以及发生的错误。
(9) 备份文件:用于存放数据库备份所产生的文件。

3.2.1 数据文件

数据文件就是用来存放数据库数据的物理文件,文件后缀为".dbf"。数据文件存放的主要内容可以归纳为表中的数据,索引数据,数据字典定义,回滚事务所需信息,存储过程、函数和数据包的代码,以及用来排序的临时数据。

数据文件可以通过动态性能视图进行查看,也可以进行很多操作。关于数据文件的具体操作将在下面具体阐述。

1. 数据文件相关视图

若要了解数据文件中的具体信息,首先需要熟悉数据文件的相关视图,具体如下所述。

(1) dba_data_files:包含数据库中所有数据文件的信息,包括数据文件所属的表空间、数据文件编号等。

(2) dba_temp_files:包含数据库中所有临时数据文件的信息。

(3) dba_extents:包含所有表空间中已分配的区的描述信息。

(4) user_extents:包含当前用户所拥有的对象在所有表空间中已分配的区的描述信息。

(5) dba_free_space:包含表空间中空闲区的描述信息。

(6) user_free_space:包含当前用户可访问的表空间中空闲区的信息。

(7) v$datafile:包含从控制文件中获取的数据文件信息。

(8) v$datafile_header:包含从数据文件头部获取的信息。

(9) v$tempfile:包含所有临时文件的基本信息。

2. 查询看数据文件信息

(1) 查询数据文件的动态信息,示例如下:

```
SQL> col name for a50
SQL> select file#,name from v$datafile;

    FILE#   NAME
 --------   ---------------------------------
        1   D:\ORACLE19C\ORADATA\NEWORCL\SYSTEM01.DBF
        3   D:\ORACLE19C\ORADATA\NEWORCL\SYSAUX01.DBF
        4   D:\ORACLE19C\ORADATA\NEWORCL\UNDOTBS01.DBF
        7   D:\ORACLE19C\ORADATA\NEWORCL\USERS01.DBF
```

(2) 查询数据文件的详细信息,示例如下:

```
SQL> col file_name for a50
SQL> col tablespace_name for a20
SQL> select file_id,file_name,tablespace_name,bytes/(1024*1024)MB
  2  from dba_data_files order by tablespace_name;

FILE_ID  FILE_NAME                                        TABLESPACE_NAME    MB
-------  -----------------------------------------------  ---------------    ----
3        D:\ORACLE19C\ORADATA\NEWORCL\SYS AUX01.DBF SYS   AUX                720
1        D:\ORACLE19C\ORADATA\NEWORCL\SYSTEM01.DBF        SYSTEM             910
4        D:\ORACLE19C\ORADATA\NEWORCL\UNDOTBS01.DBF       UNDOTBS1           65
7        D:\ORACLE19C\ORADATA\NEWORCL\USERS01.DBF         USERS              5
```

(3) 查询数据文件的增长方式,示例如下:

```
SQL> SELECT TABLESPACE_NAME,BYTES, AUTOEXTENSIBLE, FILE_NAME
  2  FROM DBA_DATA_FILES;

TABLESPACE_NAME   BYTES       AUTOEXTENSIBLE   FILE_NAME
---------------   ---------   --------------   -----------------------------------
SYSTEM            954204160   YES              D:\ORACLE19C\ORADATA\NEWORCL\SYSTEM01.DBF
SYSAUX            754974720   YES              D:\ORACLE19C\ORADATA\NEWORCL\SYSAUX01.DBF
USERS             5242880     YES              D:\ORACLE19C\ORADATA\NEWORCL\USERS01.DBF
UNDOTBS1          68157440    YES              D:\ORACLE19C\ORADATA\NEWORCL\UNDOTBS01.DBF
```

(4) 查询临时数据文件信息,示例如下:

```
SQL> select tablespace_name, file_name,
  2  autoextensible from dba_temp_files;

TABLESPACE_NAME   FILE_NAME                                      AUT
---------------   --------------------------------------------   ---
TEMP              D:\ORACLE19C\ORADATA\NEWORCL\TEMP01.DBF        YES
```

3. 重置数据文件大小

(1) 观察每个数据文件的空间大小,示例如下:

```
SQL> select tablespace_name, file_id, file_name, round (bytes /(1024 * 1024 ),0 ) total_space
  from dba_data_files order by tablespace_name;

TABLE     SPACE_NAME   FILE_IDFILE_NAME                          TOTAL_SPACE
------    ----------   ---------------------------------------   -----------
```

SYSAUX	3	D:\ORACLE19C\ORADATA\NEWORCL\SYSAUX01.DBF	720
SYSTEM	1	D:\ORACLE19C\ORADATA\NEWORCL\SYSTEM01.DBF	910
UNDOTBS1	4	D:\ORACLE19C\ORADATA\NEWORCL\UNDOTBS01.DBF	65
USERS	7	D:\ORACLE19C\ORADATA\NEWORCL\USERS01.DBF	5

（2）观察每个数据文件的使用情况，示例如下：

```
SQL> select file_name, a.file_id, sum (a.bytes )/1024 /1024 as MB
  2  from dba_extents a, dba_data_files b where a.file_id = b.file_id
  3  group by file_name, a.file_id;


FILE_NAME                                             FILE_ID        MB
---------------------------------------------------   --------   --------
D:\ORACLE19C\ORADATA\NEWORCL\USERS01.DBF                    7     1.6875
D:\ORACLE19C\ORADATA\NEWORCL\SYSTEM01.DBF                   1   900.6875
D:\ORACLE19C\ORADATA\NEWORCL\SYSAUX01.DBF                   3     645.75
D:\ORACLE19C\ORADATA\NEWORCL\UNDOTBS01.DBF                  4     32.125
```

（3）重置文件大小，示例如下：

```
SQL> alter database datafile 'D:\ORACLE19C\ORADATA\NEWORCL\SYSAUX01.DBF' resize 750MB;

数据库已更改。
```

4. 移动数据文件

若某个磁盘的 I/O 操作过于繁忙，就可能影响到 Oracle 数据库系统的整体效率，此时需将一个或几个数据文件进行移动。若某个磁盘已经毁损，为了能使数据库系统继续运行，要将一个或多个数据文件移动。移动数据文件的操作语句有如下两种。

（1）alter tabespace 表空间名 rename datafile 文件名 to 文件名。

该语句只适用于上面没有活动的还原数据或临时段的非系统表空间中的数据文件。要求在使用该语句时，表空间必须为脱机状态且目标数据文件必须存在，因为该语句只修改控制文件中指向数据文件的指针（地址）。

（2）alter database 数据库名 rename file 文件名 to 文件名。

该语句适用于系统表空间和不能置为脱机的表空间中的数据文件。要求在使用该语句时，数据库必须运行在加载（Mount）状态且目标数据文件必须存在，因为该语句只修改控制文件中指向数据文件的指针（地址）。

3.2.2 控制文件

控制文件用于记录和维护数据库的全局物理结构，它就像数据库的一个管家，是成功启动和操作数据库必须的二进制文件，以".ctl"为文件后缀。一个数据库至少需要一个控制文件，每个控制文件只与一个数据库相关联。

控制文件中包含的记录项有数据库名称和标识符、数据库创建时间、表空间名称、数据文件和联机重做日志的名字和位置、当前联机重做日志序号、检查点信息、还原段的开始与结束、重做日志归档/存档信息和备份信息。

1. 控制文件相关视图

若要进行控制文件的有关操作,首先需要了解有关控制文件的视图,如下所示。

(1) v$controlfile:列出所有与当前 Oracle 实例相关的控制文件的名和状态。
(2) v$parameter:列出所有参数的状态和位置。
(3) v$controlfile_record_section:给出控制文件记录段相关的信息。

2. 获取控制文件

通过获取控制文件,可以发现每个数据库通常包含两个或更多控制文件,这几个控制文件在内容上保持一致,也分配在相同的物理硬盘中。但当数据库或硬盘损坏时,可利用备份的控制文件启动 Oracle 实例,以提高数据库的可靠性。

```
SQL> select name from v$controlfile;
NAME
--------------------------------------------------------------
D:\ORACLE19C\ORADATA\NEWORCL\CONTROL01.CTL
D:\ORACLE19C\ORADATA\NEWORCL\CONTROL02.CTL
```

或者

```
SQL> select value from v$parameter where name = 'control_files';

VALUE
--------------------------------------------------------------
D:\ORACLE19C\ORADATA\NEWORCL\CONTROL01.CTL, D:\ORACLE19C\ORADATA\NEWORCL\CONTROL02.CTL
```

3. 备份控制文件

控制文件是一个极其重要的文件,除了将控制文件的多个副本存在不同的硬盘上这个保护措施外,在数据库的结构发生变化之后,还应该立即对控制文件进行备份。

```
SQL> alter database backup controlfile to 'D:\BACKUP\CONTROL.BAK';

数据库已更改.
```

或者

```
SQL> alter database backup controlfile to trace;

数据库已更改.
```

创建控制文件的命令并将其备份到一个跟踪文件中,获得当前跟踪文件生成路径。

```
SQL> select tracefile from v$process where addr in (select paddr from v$session where sid in
(select sid from v$mystat));

TRACEFILE
--------------------------------------------------------------------------------
D:\ORACLE19C\diag\rdbms\neworcl\neworcl\trace\neworcl_ora_35488.trc
```

4. 添加和移动控制文件

如果控制文件在同一目录下,就需要进行移动或添加。

步骤1　首先利用数据字典 v$controlfile 获取现有控制文件的名字。

```
SQL> select name from v$controlfile;

NAME
--------------------------------------------------------------------------------
D:\ORACLE19C\ORADATA\NEWORCL\CONTROL01.CTL
D:\ORACLE19C\ORADATA\NEWORCL\CONTROL02.CTL
```

步骤2　在数据库启动状态下,修改服务器端初始化参数文件 spfile,使用 alter system set control_files 命令改变控制文件的位置和新增控制文件。

```
SQL> alter system set control_files =
'D:\ORACLE19C\ORADATA\NEWORCL\CONTROL01.CTL','D:\ORACLE19C\ORADATA\NEWORCL\CONTROL02.CTL',
'D:\BACKUP\CONTROL03.CTL' SCOPE = SPFILE;

系统已更改.
```

步骤3　关闭数据库,以确保复制后的控制文件与源控制文件内容完全相同。

```
SQL> shutdown immediate;
数据库已经关闭.
已经卸载数据库.
ORACLE 例程已经关闭.
```

步骤4　移动或复制控制文件。可以使用当前的两个控制文件中的任意一个,复制为新的控制文件。

```
SQL> host copy D:\ORACLE19C\ORADATA\NEWORCL\CONTROL02.CTL
D:\BACKUP\CONTROL03.CTL
已复制         1 个文件.
```

步骤 5 启动数据库。

```
SQL> startup
ORACLE 例程已经启动.

Total System Global Area   2550136752 bytes
Fixed Size                    9031600 bytes
Variable Size               570425344 bytes
Database Buffers           1946157056 bytes
Redo Buffers                 24522752 bytes
数据库装载完毕.
数据库已经打开.
```

步骤 6 查询是否已经成功移动到所要移动的目录下。

```
SQL>select name from v$controlfile;

NAME
--------------------------------------------------
D:\ORACLE19C\ORADATA\NEWORCL\CONTROL01.CTL
D:\ORACLE19C\ORADATA\NEWORCL\CONTROL02.CTL
D:\BACKUP\CONTROL03.CTL
```

3.2.3　日志文件

日志文件用来记录数据库的事务处理过程,主要保存了用户对数据库所做的更新操作,包含的主要信息是记录事务的开始和结束、事务中每项操作的对象和类型、更新操作前后的数据值等。用户对数据库所做的修改都是在数据库的高速缓冲区中进行的,同时将产生的重做记录写入重新日志缓冲区,在一定条件下由 DBWn 进程将修改后的结果成批地写回到数据文件中,而重做日志缓冲区中的重做记录由 LGWR 进程周期性地写入重做日志文件中。因为所有的处理都记录在重做日志里,所以数据库系统可以使用这些事务记录进行恢复操作,后缀名为".log"。

为了保证 LGWR 进程的正常进行,通常采用重做日志组(Group),每个组中包含若干完全相同的重做日志文件成员,这些成员文件互为镜像。每组内的日志文件的内容完全相同,且保存在不同的位置。它们用于磁盘日志镜像,以做多次备份提高安全性。在默认情况下,多组通常只有一组处于活动状态。采用循环写的方式进行工作。当一个重做日志写满后,LGWR 进程就会移到下一个日志组,称之为日志切换,同时信息回写到控制文件中。

重做日志结构及工作原理如图 3-3 所示。

1. 重做日志相关视图

Oracle 提供了 v$log、v$logfile 和 v$log_history 3 个视图用于维护在线重做日志。这 3

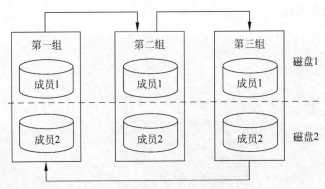

图 3-3 重做日志结构及工作原理

个视图主要用于查看和修改在线日志。

(1) v$log：包含从控制文件中获取的所有重做日志文件组的基本信息。

(2) v$logfile：包含重做日志文件组及其成员文件的信息。

(3) v$log_history：包含关于重做日志文件的历史信息。

2. 重做日志以及日志组信息查询

(1) 查询重做日志文件组的信息。

```
SQL> select group#,sequence#,members,archived,status,first_time from v$log;

   GROUP#   SEQUENCE#   MEMBERS  ARC  STATUS     FIRST_TIME
---------- ---------- ---------- ---  --------   -----------------
        1          1          1  YES  INACTIVE   09-2月 -20
        2          2          1  YES  INACTIVE   12-2月 -20
        3          3          1  NO   CURRENT    15-2月 -20
```

上述结果中 STATUS 代表日志组状态，它的值有以下 6 种。

① UNUSED：从未对联机重做日志文件组进行写入。这代表是刚添加的联机重做日志文件的状态。

② CURRENT：当前的联机重做日志文件组。这说明该联机重做日志文件组是活动的。

③ ACTIVE：联机重做日志文件组是活动的，但是并非当前联机重做日志文件组。数据库的崩溃恢复需要该组。它可能已归档，也可能未归档。

④ CLEARING：在执行 alter database clear logfile 命令后正在将该日志重建为一个空日志。日志清除后，其状态更改为 UNUSED。

⑤ CLEARING_CURRENT：正在清除当前日志文件中的已关闭线程。如果在切换时发生某些故障，如写入新日志标头时发生了输入/输出错误，那么日志可能处于此状态。

⑥ INACTIVE：例程恢复不再需要联机重做文件日志组。它可能已归档，也可能未归档。

(2) 查询重做日志文件的信息。

```
SQL>select * from v$logfile;

GROUP#    STATUS TYPE    MEMBER                                              IS_    CON_ID
------    ------------   --------------------------------------------------  ----   -------
3                ONLINE  D:\ORACLE19C\ORADATA\NEWORCL\REDO03.LOG             NO     0
2                ONLINE  D:\ORACLE19C\ORADATA\NEWORCL\REDO02.LOG             NO     0
1                ONLINE  D:\ORACLE19C\ORADATA\NEWORCL\REDO01.LOG             NO     0
```

上述结果中 STATUS 代表日志状态,它的值可以为下列值之一。

① INVALID:表明该文件不可访问。

② STALE:表示文件内容不完全。

③ DELETED:表明该文件已不再使用。

④ 空白表明文件正在使用中。

3. 强制产生日志切换

LGWR 按顺序向联机重做日志组写入重做信息。一旦当前联机重做日志组被写满,LGWR 就开始写入下一个组,称之为日志切换(Log Switch)。通常,只有当前的重做日志组在写满后才发生日志切换,并可通过设置参数 archive_log_target 来控制日志切换的时间间隔,在必要时也可以采用手工强制进行日志切换。

(1) 设置 fast_start_mttr_target:可以在初始化参数文件中设置此参数,代表实例恢复所用时间,单位为 s。例如,设置 fast_start_mttr_target=300,代表如果数据库需要实例恢复,那么恢复的时间不超过 300s,系统会根据 300s 时间自动计算可以保留的脏数据块的数目;如果恢复的时间超过 300s,那么实例会自动发出检查点。

(2) 强制产生日志切换命令。

```
SQL>alter system switch logfile;

系统已更改。
```

(3) alter system checkpoint 命令:检查点是一个数据库事件,它用于同步所有数据文件、控制文件以及重做日志文件。在必要时,DBA 也可以手动发出检查点命令,命令如下:

```
SQL>alter system checkpoint;

系统已更改。
```

4. 重做日志的管理

通常,DBA 会在创建数据库时按照计划创建所需重做日志组和各个组成员日志文件。然而在有些情况下,需要通过手工方式为数据库添加新重做日志组和成员。例如,如果当前某个重做日志组由于某种原因无法使用,DBA 需要创建一个新的重做日志组代替它进行工作。在另外一些情况下,DBA 可能需要改变现有重做日志文件的名称和位置或删除重做日志组或成员。

(1) 创建重做日志组语法：

```
alter dtabase add logfile [group n]
```

(2) 删除重做日志组语法：

```
alter database drop logfile [group n]
```

(3) 创建重做日志文件组成员文件：

```
alter database add logfile member '文件名' to group n
```

(4) 删除重做日志文件组成员文件：

```
alter database drop logfile member '文件名'
```

(5) 重新初始化联机重做日志组：

```
alter database clear logfile group n
```

(6) 清除崩溃的重做日志文件使其不能归档：

```
alter database clear unarchived logfile group n
```

(7) 当修改完相关日志文件后，可以使用数据字典 v$log 和 v$logfile 进行查看。

3.2.4 参数文件

通过学习 Oracle 数据库的体系结构，实例是一组 Oracle 后台进程和内存结构的集合，那么实例到底占用多大内存空间，并且在启动实例时是否要启动某些特定的后台进程，这都需要通过配置参数文件来完成。如以 sysdba 身份发出 startup 命令，Oracle 服务器就会读取初始化参数文件，并根据参数文件来配置实例。在启动实例时必须有相应的初始化参数存在。数据库参数文件主要用于保存数据库的非默认参数。

参数文件如图 3-4 所示。

1. 参数文件分类

Oracle 有两种不同类型的初始化参数文件。

(1) 静态参数文件(pfile)：该文件为正文文件。静态参数文件的文件名一般为 initSID.ora。

(2) 动态服务器参数文件(spfile)：该文件为二进制文件。动态服务器参数文件的文件名一般为 spfileSID.ora。这里的 SID 为实例名。

2. 查看参数文件

为了提高 Oracle 服务器运行性能，可能需要经常查询或修改相关参数。那么如何监视

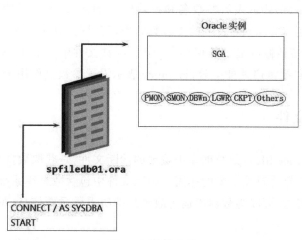

图 3-4 参数文件

初始化参数的设置呢?可通过如下方法实现。

(1) show parameter:显示当前会话中所有初始化参数的值。

```
SQL> show parameter pfile;

NAME          TYPE       VALUE
-----------   --------   ---------------------------------------
spfile        string     D:\SOFTWARE\ORACLE19C\DATABASE\SPFILENEWORCL.ORA
SQL> show parameter spfile;

NAME          TYPE       VALUE
-----------   --------   ---------------------------------------
spfile        string     D:\SOFTWARE\ORACLE19C\DATABASE\SPFILENEWORCL.ORA
```

(2) v$paramenter:包含当前会话中所有初始化参数及其值。

```
SQL> select name,value from v$parameter where name = 'spfile';

NAME      VALUE
------    ---------------------------------------
spfile    D:\SOFTWARE\ORACLE19C\DATABASE\SPFILENEWORCL.ORA
```

3. 修改参数

修改 Oracle 参数的 SQL 语法为:

```
alter system  set parameter = value
<deferred>
<scope = both|spfile|memory>
<sid = 'sid| * '>
```

其中：parameter＝value 表示参数名和参数值；deferred 表示设置参数对当前会话不生效,对以后的会话生效；scope 表示设置参数的作用范围。其中,both 表示设置参数在当前实例中生效,并将参数修改后的值保存在 spfile 参数文件中；spfile 表示设置参数仅保存在 spfile 参数文件中,要重启才能生效；memory 表示设置参数仅作用于当前实例。

3.2.5 其他文件

（1）跟踪文件(trace file)：是数据库中重要的诊断文件,是获取数据库信息的重要工具,对管理数据库的实例起着至关重要的作用。跟踪文件中包含数据库系统运行过程中所发生的重大事件的有关信息,可以为数据库运行故障的解决提供重要信息,每个后台进程都有相应的跟踪文件。

（2）告警文件(alert file)：是数据库中重要的诊断文件,记录数据库在启动、关闭和运行期间后台进程的活动情况。其中跟踪文件与告警文件将在后续章节详细展开讨论。

（3）备份文件(backup file)：是历史联机重做日志文件的集合,是联机重做日志文件被覆盖之前备份的副本。

（4）口令文件(password file)：存放用户口令的加密文件。

3.3 Oracle 启动与关闭

各文件之间的关系如图 3-5 所示。从图中可以发现,文件之间与数据库的启动息息相关。

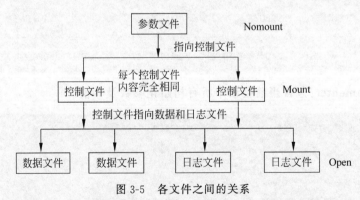

图 3-5 各文件之间的关系

由图 3-5 所示的各文件之间的关系可引入 Oracle 的物理结构,建立层次的概念,同时引入数据库启动的 3 个阶段。在讲解这部分内容的时候,一定要结合构成层次结构的 3 个层次的文件和启动过程的关系展开,从而将这些基本的概念统合到一起。

Oracle 的启动分为 3 个阶段：启动实例(Nomount)、装载数据库(Mount)和打开数据库(Open)。

Oracle 的正常关闭也分为 3 个阶段：关闭数据库(Closed)、卸载数据库(Dismounted)和终止进程(Instance Shutdown)。

下面具体了解一下 Oracle 的启动和关闭。

3.3.1　Oracle 的启动过程

Oracle 数据库的启动分为 3 个阶段：启动实例、装载数据库和打开数据库。这里的每个阶段都会读取和校验不同的数据库文件，并按照时间的先后顺序将相关信息写入到告警日志中，所以在启动等过程中如果出现问题，就可以观察和研究告警日志。这也是学习 Oracle 数据库的一个重要途径。

1. 启动实例（Nomount）

Oracle 会读取一个参数文件（PFILE 或者 SPFILE 文件），Oracle 根据参数文件中的参数，分配一系列后台进程和服务进程，并且在内存中创建 SGA 区等内存结构。内存和进程就组成了所谓的实例。每一个进程都拥有一个自己的名字（SID）。在实例启动完毕，数据库还没有跟实例关联，还处于 Nomount 状态，表明数据库还不可以访问。

实例启动的语句为：

```
SQL> startup nomount
```

在 Nomount 模式下，只能访问那些与 SGA 区相关的数据字典视图，包括 v$parameter、v$sga、v$process 和 v$session 等，这些视图中的信息都是从 SGA 区中获取的，与数据库无关。实例启动可执行重建控制文件、重建数据库，读取 init.ora 文件，启动 instance，即启动 SGA 和后台进程。这种启动只需要 init.ora 文件。

实例启动过程如图 3-6 所示。

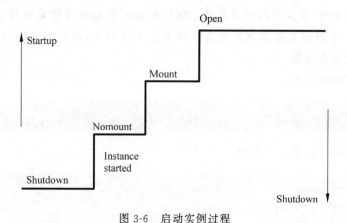

图 3-6　启动实例过程

2. 装载数据库（Mount）

装载阶段创始实例并且安装数据库，但没有打开数据库。Oracle 系统读取控制文件中关于数据文件和重做日志文件的内容，但是并不打开该文件。在这种打开方式下，系统会给出"数据库装载完毕"的提示。

装载数据库的语句为：

```
SQL> startup mount
```

这种启动模式将为实例加载数据库,但保持数据库为关闭状态。因为加载数据库时需要打开数据库控制文件,但数据文件和重做日志文件都无法进行读写,所以用户还无法对数据库进行操作。在 Mount 模式下,只能访问那些与控制文件相关的数据字典视图,包括 v$thread、v$controlfile、v$database、v$datafile 和 v$logfile 等,这些视图都是从控制文件中获取的。这种打开方式经常在数据库维护操作时使用,如数据库日志归档、数据库介质恢复、使数据文件联机或脱机、重新定位数据文件和重做日志文件。

装载数据库过程如图 3-7 所示。

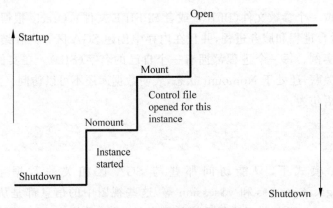

图 3-7 装载数据库过程

3. 打开数据库(Open)

在打开数据库时,实例将打开所有处于联机状态的数据文件和重做日志文件。若控制文件中的任何一个数据文件或重做日志文件无法正常打开,数据库都将会返回错误信息。这时需要进行数据库恢复。

打开数据库的语句为:

```
SQL> startup [open]
```

这个命令等同于以下 3 个命令。

```
SQL> startup nomount;
SQL> alter database mount;
SQL> alter database open;
```

只有设置为打开状态,数据库才处于正常状态,这时普通用户才能够访问数据库。在很多情况下,启动数据库时并不是直接完成上述 3 个步骤,而是逐步完成的,然后执行必要的管理操作,最后才使数据库进入正常运行状态。所以才有了各种不同的启动模式用于不同的数据库维护操作。

打开数据库过程如图 3-8 所示。

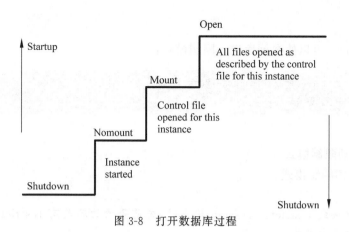

图 3-8　打开数据库过程

3.3.2　数据库的启动

因为在 Oracle 数据库启动过程中,不同的阶段可以对数据库进行不同的维护操作,对应不同的需求,所以需要不同的模式启动数据库。

1. 启动命令

上面已经简单说明了实例启动的 3 个阶段的通用启动命令,下面对启动命令语法进行详细讲解。

```
startup [force][restrict][pfile = filename]
[open [recover] [dbname]
|mount|nomount]
```

其中:

open:启动实例,装载并打开数据库,为默认选项;

mount:启动实例并装载数据库,但不打开数据库;

nomount:启动实例,但不能装载数据库;

pfile=:指定用于启动实例的非默认初始化参数文件名;

force:强制中止实例,并重新启动数据库;

restrict:启动后只允许具有 restricted session 权限的用户访问数据库;

recover:在数据库启动时进行介质恢复。

2. 只读状态打开数据库

在正常启动状态,默认数据库进入读/写状态,在必要时可以将数据库设置为只读状态。在只读状态下,用户只能查询数据库,但不能以任何方式对数据库对象进行修改。

使用如下命令,可以使数据库进入只读状态或进入读/写状态:

```
语法 1:startup open [read write|read only]
语法 2:alter database open [read write|read only]
```

3. 用限制模式

限制模式只允许具有 restricted session 权限的用户正常使用数据库;而其他用户被限

制使用。

使用如下命令,可以使数据库进入限制模式:

```
语法 1:alter system [{enable|disable}restricted session]
语法 2:startup restrict
```

其中:
enable:启动限制模式;
disable:取消限制模式。
提示:
当数据库切换到 restricted session 状态时,先前登录的不具有 restricted session 权限的用户仍然可以正常工作。

3.3.3 数据库的关闭过程

与数据库启动一样,关闭数据库实例也需要分 3 步:关闭数据库、卸载数据库和终止进程。

1. 关闭数据库

在该阶段,Oracle 将重做日志高速缓冲区中的内容写入重做日志文件,并且将数据库高速缓冲中被改动过的数据写入数据文件,然后再关闭所有的数据文件和重做日志文件,这时数据库的控制文件仍然处于打开状态。但是由于数据库处于关闭状态,所以用户无法访问数据库。

2. 卸载数据库

在关闭数据库后,例程才能被卸载,控制文件在这个时候被关闭,但是例程仍然存在。

3. 终止进程

进程终止,分配给例程的内存 SGA 区被回收。

3.3.4 数据库的关闭

当 DBA 要执行完数据库备份、修改初始化参数以及其他系统维护操作时,需要停止 Oracle 服务器。

数据库关闭分为两类共 4 种方式:Normal(正常关闭)、Immediate(立即关闭)、Transaction(直接关闭)、Shutdown Abort(终止关闭)。

其中,前 3 种关闭方式属于一致性数据库关闭,特点是无须进行数据库恢复;而 Shutdown Abort 关闭方式属于非一致性数据库关闭。使用 Abort 关闭、数据库发生实例故障(如断电)、或使用 Startup Force 强制重新启动数据库,都需要进行实例恢复。

一致性数据库关闭与非一致性数据库关闭如图 3-9 和图 3-10 所示。

1. Normal(正常关闭)

当使用正常方式关闭数据时,Oracle 执行如下操作。

(1) 阻止任何用户建立新的连接。

(2) 等待当前所有正在连接的用户主动断开连接(此方式下 Oracle 不会立即断掉当前

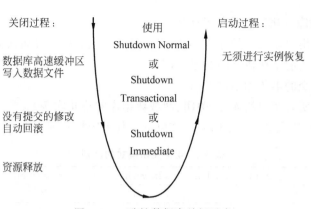

图 3-9　一致性数据库关闭示意

图 3-10　非一致性数据库关闭示意

用户的连接,这些用户仍然可操作相关的操作)。

(3) 一旦所有的用户都断开连接,就立即关闭、卸载数据库,并终止实例。

2．Immediate(立即关闭)

当采用立即关闭数据时,Oracle 执行如下操作。

(1) 阻止任何用户建立新的连接,同时阻止当前连接的用户开始任何新的事务。

(2) Oracle 不等待在线用户主动断开连接,强制终止用户的当前事务,将任何未提交的事务回退(如果存在太多未提交的事务,此方式将会耗费很长时间终止和回退事务)。

3．Transaction(直接关闭、卸载数据库,并终止实例)

这种方式介于正常关闭方式跟立即关闭方式之间,响应时间会比较快,处理也将比较得当。执行过程如下所述。

(1) 阻止任何用户建立新的连接,同时阻止当前连接的用户开始任何新的事务。

(2) 等待所有未提交的活动事务提交完毕,然后立即断开用户的连接。

(3) 直接关闭、卸载数据库,并终止实例。

4．Shutdown Abort(终止关闭)

这是比较粗暴的一种关闭方式,当前面 3 种方式都无法关闭时,可以尝试使用终止方式来关闭数据库。但是以这种方式关闭数据库将会丢失一部分数据信息,当重新启动实例并打开数据库时,后台进程 SMON 会执行实例恢复操作。一般情况下,应当尽量避免使用这种方式来关闭数据库。执行过程如下所述。

(1) 阻止任何用户建立新的连接,同时阻止当前连接的用户开始任何新的事务。

(2) 立即终止当前正在执行的 SQL 语句。

(3) 任何未提交的事务均不被退回。

(4) 直接断开所有用户的连接,关闭、卸载数据库并终止实例。

综上所述,可利用表 3-1 所示的来总结以下 4 种关闭模式的区别。

表 3-1　4 种关闭模式的区别

关闭模式	Shutdown Abort	Immediate	Transaction	Normal
允许新的连接	×	×	×	×
等待当前会话结束	×	×	×	√
等待当前事务结束	×	×	√	√
强制检查点并关闭文件	×	√	√	√

3.4　Oracle 逻辑存储

逻辑存储结构是从逻辑的角度来分析数据库的构成,是数据库创建后利用逻辑概念来描述 Oracle 数据库内部数据的组织和管理形式。在操作系统中,没有数据库逻辑存储结构信息,只有物理存储结构信息。数据库的逻辑存储结构信息存储在数据库的数据字典中,可以通过数据字典进行查询。

Oracle 数据库的逻辑结构是一种层次结构,主要由表空间、段、区和块等概念组成。逻辑结构是面向用户的,用户使用 Oracle 开发应用程序使用的就是逻辑结构。

Oracle 数据库逻辑存储结构之间的关系如图 3-11 所示。

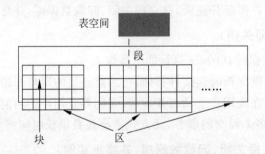

图 3-11　Oracle 数据库逻辑存储结构之间的关系

(1) 块(Block):数据库中最小的 I/O 单元。

(2) 区(Extent):由若干连续的数据块组成,是数据库中最小的存储分配单元。

(3) 段(Segment):由若干区组成,存储相同类型数据。

(4) 表空间(Tablespace):由若干段组成,是最大的存储逻辑单元。所有表空间构成数据库。

3.4.1　表空间

1. 表空间的概念

表空间(Tablespaces)是数据库的逻辑划分,一个表空间只能属于一个数据库。所有的数据库对象都存放在指定的表空间中。由于主要存放的对象是表,所以称作表空间。

Oracle 数据库的每个表空间包含一个或者多个".dbf"的操作系统文件,称为数据文件。数据文件的大小决定了表空间的大小。一个数据文件只能从属于一个表空间。

表空间是存储模式对象的容器,一个数据对象只能存储在一个表空间中(分区表和分区索引除外),但可以存储在该表空间所对应的一个或者多个数据文件中。若表空间只有一个数据文件,则该表空间中所有对象都保存在该文件中;若表空间对应多个数据文件,则表空间中的对象可以分布于不同的数据文件中。

表空间与数据文件、数据对象的关系如图 3-12 所示。

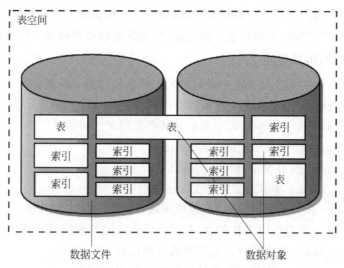

图 3-12　表空间与数据文件、数据对象的关系

表空间从逻辑上是多个段的结合,在物理上是多个数据文件的集合,相当于在段和数据文件的对应中加入了一个中间层来解决这种多对多的关系。

表空间、段、区、块之间的逻辑关系如图 3-13 所示。

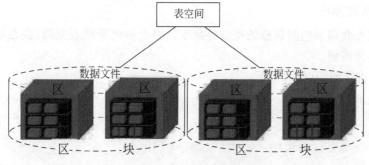

图 3-13　表空间、段、区、块之间的逻辑关系

2. 表空间的分类

数据库在安装完毕,会创建 5 个基本的表空间。关于 Oracle 表空间的分类如表 3-2 所示。

表 3-2　Oracle 表空间的分类

Tablespace 类型	用途说明
SYSTEM	存储系统数据字典的相关表、视图
SYSAUX	存放附加的数据库组件,如企业管理器资料库
TEMP	存放系统处理排序时所用的过渡型数据
UNDOTBS	存放系统执行交易回滚所需的建议前数据
USERS	用于存放用户私有信息

表空间可分为系统表空间和非系统表空间。系统表空间(包括 SYSTEM 和 SYSAUX 表空间)与数据库一起建立,为强制性表空间,必须为联机状态。而非系统(Non-SYSTEM)表空间存储一些单独的段,方便磁盘空间管理,控制分配给用户磁盘空间的数量。

3. 与表空间的相关视图

(1) v$tablespace:从控制文件中获取的表空间名称和编号信息。
(2) dba_tablespaces:数据库中所有表空间的信息。
(3) dba_tablespace_groups:表空间组及其包含的表空间信息。
(4) dba_segments:所有表空间中段的信息。
(5) dba_extents:所有表空间中区的信息。
(6) dba_free_space:所有表空间中空闲区的信息。
(7) v$datafile:所有数据文件信息,包括所属表空间的名称和编号。
(8) v$tempfile:所有临时文件信息,包括所属表空间的名称和编号。
(9) dba_data_files:数据文件及其所属表空间信息。
(10) dba_temp_files:临时文件及其所属表空间信息。
(11) dba_users:所有用户的默认表空间和临时表空间信息。
(12) dba_ts_quotas:所有用户的表空间配额信息。
(13) v$sort_segment:数据库实例的每个排序段信息。
(14) v$sort_user:用户使用临时排序段信息。

4. 表空间的操作

下面将重点介绍表空间信息的查询的命令。关于如何管理表空间,会在后续的空间管理章节展开具体讲解。

(1) 查看表空间,其代码如下:

```
SQL> select * from v$tablespace;

    TS#  NAME        INC BIG FLA ENC   CON_ID
    ---- ----------- --- --- --- ---   --------
    1    SYSAUX      YES NO  YES       0
    0    SYSTEM      YES NO  YES       0
    2    UNDOTBS1    YES NO  YES       0
    4    USERS       YES NO  YES       0
    3    TEMP        NO  NO  YES       0
```

(2) 查看详细数据文件，其代码如下：

```
SQL> col tablespace_name format a10
SQL> select file_name,tablespace_name from dba_data_files;

FILE_NAME                                          TABLESPACE
-------------------------------------------------- ----------
D:\ORACLE19C\ORADATA\NEWORCL\SYSTEM01.DBF          SYSTEM
D:\ORACLE19C\ORADATA\NEWORCL\SYSAUX01.DBF          SYSAUX
D:\ORACLE19C\ORADATA\NEWORCL\USERS01.DBF           USERS
D:\ORACLE19C\ORADATA\NEWORCL\UNDOTBS01.DBF         UNDOTBS1
```

(3) 查询表空间基本信息，其代码如下：

```
SQL> select tablespace_name,exetnt_management,allocation_type,contents
  2  from dba_tablespaces;

TABLESPACE  EXTENT_MAN  ALLOCATION  CONTENTS
----------  ----------  ----------  -------------------------
SYSTEM      LOCAL       SYSTEM      PERMANENT
SYSAUX      LOCAL       SYSTEM      PERMANENT
UNDOTBS1    LOCAL       SYSTEM      UNDO
TEMP        LOCAL       UNIFORM     TEMPORARY
USERS       LOCAL       SYSTEM      PERMANENT
```

(4) 查询表空间数据文件信息，其代码如下：

```
SQL> select file_name,blocks,tablespace_name
  2  from dba_data_files;

FILE_NAME                                          BLOCKS   TABLESPACE
-------------------------------------------------- -------  ----------
D:\ORACLE19C\ORADATA\NEWORCL\SYSTEM01.DBF          116480   SYSTEM
D:\ORACLE19C\ORADATA\NEWORCL\SYSAUX01.DBF          92160    SYSAUX
D:\ORACLE19C\ORADATA\NEWORCL\USERS01.DBF           640      USERS
D:\ORACLE19C\ORADATA\NEWORCL\UNDOTBS01.DBF         8320     UNDOTBS1
```

(5) 查询表空间空闲空间大小，其代码如下：

```
SQL> select tablespace_name,sum(bytes)free_spaces
  2  from dba_free_space
  3  group by tablespace_name;
```

```
TABLESPACE                      FREE_SPACES
------------------------------  ------------------------------
SYSTEM                          8716288
SYSAUX                          71892992
UNDOTBS                         131850496
USERS                           2424832
```

(6) 统计表空间空闲空间信息,其代码如下:

```
SQL> select tablespace_name "tablespace", file_id, count(*)"fieces", max(blocks) "maximum",
  2    min(blocks) "minimun", avg(blocks) "avgerage", sum(blocks) "total"
  3  from dba_free_space
  4  group by tablespace_name, file_id;

TABLESPACE      FILE_ID    FIECES    MAXIMUM    MINIMUN    AVGERAGE       TOTAL
----------      -------    ------    -------    -------    --------       -----
SYSAUX          3          117       4736       8          75.008547      8776
UNDOTBS         14         8         3328       16         486            3888
SYSTEM          1          2         1024       40         532            1064
USERS           7          1         296        296        296            296
```

3.4.2 段

段(Segment)是由一个或多个连续或不连续的区组成的逻辑存储单元,用于存储特定的、具有独立存储结构的数据库对象。一般情况下,一个数据库对象拥有一个段。

根据存储对象类型的不同,可将段分为表段、索引段、临时段和还原段 4 类。

(1) 表段又称数据段,用来存储表或簇的数据,可以细分为普通表段(Table)、分区表段(Table Partition)、簇段(Cluster)、索引化表段(Index-Organized Table)。

(2) 索引段(Index Segment)用来存放索引数据,包括 rowid 和索引值。

(3) 临时段(Temporary Segment)是在进行查询、排序等操作时,如果内存空间不足,用于保存 SQL 语句在解释和执行过程中产生的临时数据。在会话结束时,为该操作分配的临时段将被释放。

(4) 还原段(Undo Segment)用于保存数据库的回退信息,包含当前未提交事务所修改的数据的原始版本。利用回退段中保存的回退信息,可以实现事务回滚、数据库恢复、数据的读一致性和闪回查询。

3.4.3 区

区(Extent)是数据库存储空间分配的一个逻辑单位,Oracle 数据库在分配空间时,并不是以块为单位进行的,而是多个连续的块一次性地分配给数据库对象。这些连续的块就是区。

当创建一个数据库对象时,Oracle 数据库为这些对象创建一个段,并分配初始区。当段中的初始区的存储空间使用完毕后,Oracle 数据库会为段自动分配新的区,且每个区的大小不要求相同。

3.4.4 块

1. 块的定义

块(BLock)是 Oracle 数据库管理数据文件中存储空间的单位,为数据库使用的 I/O 的最小单位,其大小可不同于操作系统的标准 I/O 块大小。

在 Oracle 19c 数据库中,块分为标准块和非标准块两种。其中,标准块由数据库初始化参数 db_block_size 设置,其大小不可更改。Oracle 数据库的默认数据缓冲区就是标准块构成的。块尺寸是处理 Oracle 数据库更新、选择或者插入数据事务的最小单位。通过下列查询,可观察默认标准块的大小为 8192B,即 8KB。

```
SQL> show parameter db_block_size

NAME                                 TYPE        VALUE
------------------------------------ ----------- ----------
db_block_size                        integer     8192
SQL> select value from v$parameter where name = 'db_block_size';
VALUE
--------------------------------------------------
8192
```

2. 块的结构

Oracle 中块的结构如图 3-14 所示。

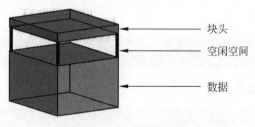

图 3-14　块的结构示意

块由块头、数据区和空闲区 3 个部分组成。其中,上面是块的头部分,下面是块的数据部分,而中间为空闲区。头部从上往下增长,而数据部分则从下往上增长,当两部分接触时块就满了。

在块中各个区域的存储信息如图 3-15 所示。

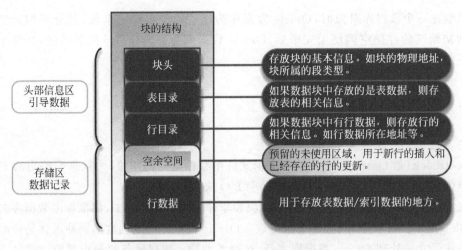

图 3-15 在块中各个区域的存储信息

3.5 项目案例

3.5.1 内存结构实例

视频讲解

通过以下案例来验证 SGA 中 database buffers 和 shared pool 的作用,具体的操作步骤如下所示。

步骤 1 设定 SQL Plus 显示时间(注意,区别于 set time on)。

```
SQL > set timing on;
```

步骤 2 第一次执行。

```
SQL > select count( * ) from dba_objects;
```

步骤 3 第二次执行。

```
SQL > select count( * ) from dba_objects;
```

步骤 4 清空共享缓存。

```
SQL > alter system flush shared_pool
```

步骤 5 清空数据库块缓存。

```
SQL > alter system flush buffer_cache
```

3.5.2 日志文件操作实例

该实验以日志文件为例,介绍日志文件的基本操作,包括创建/删除日志文件组、添加/

删除日志文件、查看日志文件/文件组、强制日志切换等内容。

步骤 1　删除重做日志文件组 3。

```
SQL> alter database drop logfile group 3;
```

数据库已更改.

步骤 2　向文件组 2 中添加一个日志文件。

```
SQL> alter database add logfile member
'D:\ORACLE19C\ORADATA\NEWORCL\REDO02-2.LOG' to group 2;
```

数据库已更改.

步骤 3　添加重做日志文件组 3,该文件组由 3 个日志文件组成。

```
SQL> alter database add logfile group 3
('D:\ORACLE19C\ORADATA\NEWORCL\REDO03-1.LOG','D:\ORACLE19C\ORADATA\NEWORCL\REDO03-2.LOG','D:\ORACLE19C\ORADATA\NEWORCL\REDO03-3.LOG') size 15MB;
```

数据库已更改.

步骤 4　查看当前日志情况。

```
SQL> select group#,sequence#,members,archived,status,first_time from v$log;

    GROUP#  SEQUENCE#    MEMBERS ARC STATUS       FIRST_TIME
---------- ---------- ---------- --- ------------ ----------------
         1          4          1 NO  CURRENT      15-2月 -20
         2          2          2 YES INACTIVE     12-2月 -20
         3          0          3 YES UNUSED
```

步骤 5　进行重做日志文件的切换。

```
SQL> alter system switch logfile;
```

系统已更改.

步骤 6　观察日志文件组状态的变化。

```
SQL> select group#,sequence#,members,archived,status,first_time from v$log;

    GROUP#  SEQUENCE#    MEMBERS ARC STATUS       FIRST_TIME
---------- ---------- ---------- --- ------------ ----------------
         1          4          1 YES ACTIVE       15-2月 -20
         2          2          2 YES INACTIVE     12-2月 -20
         3          5          3 NO  CURRENT      15-2月 -20
```

步骤 7 观察日志文件状态的变化。

```
SQL> select * from v$logfile;

   GROUP#  STATUS    TYPE   MEMBER                                          IS_     CON_ID
   ------  -------   ----   ---------------------------------------------   -----   ------
        2  INVALID   ONLINE D:\ORACLE19C\ORADATA\NEWORCL\REDO02-2.LOG       NO           0
        2            ONLINE D:\ORACLE19C\ORADATA\NEWORCL\REDO02.LOG         NO           0
        1            ONLINE D:\ORACLE19C\ORADATA\NEWORCL\REDO01.LOG         NO           0
        3            ONLINE D:\ORACLE19C\ORADATA\NEWORCL\REDO03-1.LOG       NO           0
        3            ONLINE D:\ORACLE19C\ORADATA\NEWORCL\REDO03-2.LOG       NO           0
        3            ONLINE D:\ORACLE19C\ORADATA\NEWORCL\REDO03-3.LOG       NO           0

已选择 6 行。
```

3.5.3 参数文件操作实例

视频讲解

下面以 log_buffer 参数为例,通过实验来介绍修改该参数的具体操作步骤。

步骤 1 查看该参数的值。

```
SQL> show parameter log_buffer;

NAME                                 TYPE        VALUE
-----------------------------------  ----------- ----------------------
log_buffer                           big integer 10MB
SQL> select name,value,display_value from v$parameter where name = 'log_buffer';
NAME        VALUE                                              DISPLAY_VALUE
---------   ------------------------------------------------   --------------
log_buffer  10485760                                           10MB
```

步骤 2 修改该参数的值。

```
SQL> alter system set log_buffer = 7232KB scope = spfile;

系统已更改。
```

步骤 3 验证修改后的值。

```
SQL> show parameter log_buffer;

NAME                                 TYPE        VALUE
-----------------------------------  ----------- ----------------------
log_buffer                           big integer 10MB
```

该参数的值未修改的原因是作用范围为 spfile,必须重启系统后才能看到调整后的结果。

3.5.4 数据库打开与关闭实例

通过下面实验来验证数据库的打开流程以及多种关闭方式。

1. Shutdown Immediate 关闭模式测试步骤

步骤 1 在第一个用户进程命令窗口中启动数据库到 Open 阶段,以用户 hr 登录,在该用户中创建表 t1,并插入数据未提交。

视频讲解

```
SQL> conn hr
输入口令:
已连接.
SQL> create table t1(a int);
表已创建.
SQL> insert into t1   values(1);
已创建 1 行.
```

步骤 2 打开第二个用户进程命令窗口,以用户 sys 登录并关闭数据库 Shutdown Immediate。

```
C:\WINDOWS\system32>set oracle_sid=neworcl
C:\WINDOWS\system32>sqlplus
SQL PLUS: Release 19.0.0.0.0 - Production on 星期六 2月 15 19:45:33 2020
Version 19.3.0.0.0
Copyright (c) 1982, 2019, Oracle.   All rights reserved.
请输入用户名:  sys as sysdba
输入口令:
连接到:
Oracle Database 19c Enterprise Edition Release 19.0.0.0.0 - Production
Version 19.3.0.0.0

SQL> shutdown immediate
数据库已经关闭.
已经卸载数据库.
ORACLE 例程已经关闭.
```

步骤 3 重新启动数据库,查询表 t1 数据的情况,发现已经丢失。

```
SQL> startup
ORACLE 例程已经启动.


Total System Global Area 2550136752 bytes
```

```
Fixed Size                    9031600 bytes
Variable Size               570425344 bytes
Database Buffers           1962934272 bytes
Redo Buffers                  7745536 bytes
数据库装载完毕.
数据库已经打开.
SQL> select * from hr.t1;
未选定行
```

2. Shutdown Abort 关闭模式测试步骤

步骤 1　打开第一个 Command 窗口,以用户 hr 登录,向表 t1 中插入数据并提交。

```
SQL> conn hr
输入口令:
已连接.
SQL> insert into t1  values(10);
已创建 1 行.
SQL> commit;
提交完成.
```

步骤 2　打开第二个 Command 窗口,以用户 hr 登录,插入数据未提交。

```
SQL> conn hr
输入口令:
已连接.
SQL> insert into t1  values(20);
已创建 1 行.
```

步骤 3　打开第三个 Command 窗口,以用户 sys 登录,运行 Shutdown Abort 命令关闭数据库,此时会跳过前两个阶段并直接关闭实例。

```
SQL> shutdown abort;
Oracle 例程已经关闭.
```

步骤 4　重启第四个 Command 窗口,以用户 hr 登录,查看表 t1 数据情况。查询结果显示,最初插入的数据 10 存在,后续插入的数据 20 不在。这说明数据 10 即使没来得及从缓存中存入数据文件,commit 也已经将修改记录在 redo log 中,所以在重启过程中进行实例恢复,不会丢失数据。

```
SQL> startup
Oracle 例程已经启动.

Total System Global Area 2550136752 bytes
```

```
Fixed Size                  9031600 bytes
Variable Size             570425344 bytes
Database Buffers         1962934272 bytes
Redo Buffers                7745536 bytes
数据库装载完毕.
数据库已经打开.
SQL> conn hr
输入口令:
已连接.
SQL> select * from hr.t1;
         A
----------
        10
```

其中,在重启的过程中,系统会进行实例恢复,恢复的过程全部记录在告警文件中。在后面章节中会具体介绍告警文件和跟踪文件。

第4章 数据字典与动态性能视图

PPT 视频讲解

4.1 静态数据字典

数据字典是每个 Oracle 数据库的核心,它存储了非常重要的控制信息,这些信息描述了数据库本身和数据库中的各种对象,它们是以只读表和视图的形式存放在表空间 SYSTEM 中,这些对象是由用户 sys 所拥有,由 Oracle 服务器来管理的。

大家应该熟练掌握常用的数据字典,因为数据字典提供的信息很多,如:数据库的逻辑结构和物理结构;数据库对象的定义和空间分配;数据完整性约束;用户信息;角色信息;权限信息;审计信息。

只读表中的数据字典保存了对数据库的描述,其是在执行 create database 命令时创建的,创建的脚本是在< Oracle_HOME >\RDBMS\ADMIN\SQL.BSQ 中。由于基表是非标准的,并且数据是以加密形式存储的,用户通常不会直接访问基表。

本章中提到的数据字典指的是数据字典视图。数据字典视图是静态的,是指其中的内容保存在磁盘上,相对不会发生频繁的变动。数据字典视图简化了基表的信息并展现给用户,通过公共的别名去访问,由数据库脚本< Oracle_HOME >\RDBMS\ADMIN\CATALOG.SQL 创建。

数据字典视图可分为 3 类,分别由 3 个前缀标志,如下:

(1) dba_视图:数据库中的所有视图。其只有 DBA 和具有相关权限的用户才能访问。

(2) all_视图:存储的是当前用户可以访问的对象(当前用户不需要拥有它)。

(3) user_视图:存储的是当前用户所拥有和可以访问的对象。

这 3 者之间存储的数据是有重叠的,只是访问范围不同而已。数据字典之间的关系图如图 4-1 所示。

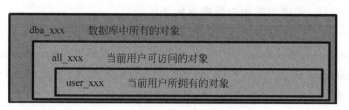

图 4-1 数据字典之间的关系图

4.1.1 通用概要类型的数据字典视图

数据字典视图包含很多类型,如通用概要类型、模式对象、磁盘空间分配、数据库的结构等。其中,通用概要类型的数据字典视图有 dictionary 和 dict_columns 两个。它们保存了当前数据库中的所有数据字典视图的信息,包括后面会介绍的动态性能视图。

1. dictionary(v$fixed_table)

dictionary 是全部数据字典的名称和解释,它有一个同义词 dict,示例代码如下:

```
SQL> col table_name format a20
SQL> col comments format a60
SQL> select * from dict where table_name = 'DBA_TABLES';


TABLE_NAME           COMMENTS
-------------------- ------------------------------------------------------------
DBA_TABLES           Description of all relational tables in the database
```

2. dict_columns

dict_columns 是全部数据字典里字段名称和解释,示例代码如下:

```
SQL> col comments format a50
SQL> select * from dict_columns
  2  where table_name = 'DBA_TABLES' and column_name = 'INITIAL_EXTENT';


TABLE_NAME COLUMN_NAME                    COMMENTS
---------- ------------------------------ --------------------------------------
DBA_TABLES INITIAL_EXTENT                 Size of the initial extent in bytes
```

4.1.2 常用的数据字典视图

下面以 user_ 前缀的视图为例介绍一些常用的数据字典视图。
(1) user_users:描述用户的信息,包括用户名、账户 ID、账户状态、表空间名等。
(2) user_tablespaces:描述当前用户可以访问的表空间。

(3) user_tables：描述当前用户所拥有的表的信息。

(4) user_views：当前用户所拥有的视图的信息。

(5) user_objects：描述当前用户所有对象的信息，包括 sequence、procedure、database link、package、package body、type body、trigger、materialized view、dimension、index、table、synonym、view、function、type 等类型。

(6) user_tab_privs：存储当前用户下对所有表的权限信息。

(7) user_errors：存储了在当前用户所拥有的对象中所发生的错误。

(8) user_source：包含了系统中对象的源码。

其实，很多对象如 db_link、extents、index、job、sequence、segment 等都可以通过 user_<对象名复数> 的表名来访问。数据字典中各个字段的详细意义在 *Oracle Database Reference* 中都有详细介绍。*Oracle Database Reference* 是非常重要的一份文档，不仅包括数据字典的解释，还包括参数文件中各个参数的详细意义以及动态性能视图等信息。

下面以最常用的关于表模式的 3 个静态视图来查看 3 种视图结构的不同。

步骤 1　user_tables：用于显示当前用户所拥有的所有表，它只返回用户所对应方案的所有表。以用户 hr 登录，查询该用户下的所有表。

```
SQL> conn hr
输入口令:
已连接.
SQL> select table_name from user_tables;

TABLE_NAME
----------------------
REGIONS
LOCATIONS
DEPARTMENTS
JOBS
EMPLOYEES
JOB_HISTORY
T1
EMP_DUMP
TEST1
COUNTRIES
SYS_TEMP_FBT

已选择 11 行.
```

步骤 2　all_tables：用于显示当前用户可以访问的所有表，它不仅会返回当前用户方案的所有表，还会返回当前用户可以访问的其他方案的表。以用户 hr 登录查询当前用户方案下所有的表。

```
SQL> select table_name from all_tables;

TABLE_NAME
----------------------
ICOL $
COL $
IND $
TAB $
CLU $
LOB $
……
已选择 2202 行.
```

步骤 3 dba_tables：它会显示所有方案拥有的数据库表。但是若要查询这种数据库字典视图,则要求用户必须是 dba 角色或是有 select any table 系统权限。

以用户 hr 登录,执行 select * from dba_users 的结构与上面一样,是因为用户 hr 没有 DBA 的权限。但是当用户 sys 查询数据字典视图 dba_tables 时,会返回 system、sys、hr 等方案所对应的数据库表。

4.2 动态性能视图

动态性能视图是动态的,不保存在磁盘上,而是存在内存中的一些虚表。其主要用于记录当前数据库的活动情况。说它是动态的,是因为在数据库运行的同时,它的内容会根据数据库的运行状态而发生频繁变化,用于监控和调整数据库,同数据字典视图一样,它也是由 sys 用户拥有。总的来说,它们均是以 v$、gv$、x $ 开头的一些同义词。

1. 动态性能视图的总览视图

相对于数据字典中的 dictionary 视图一样,动态性能视图中也有一张视图保持了所有动态性能视图的总览视图 v$fixed_table；另一张视图 v$fixed_view_definition 包含了对这些动态性能视图的定义。

```
SQL> col name format a30
SQL> select name, object_id, type, table_num from v$fixed_table;

NAME                           OBJECT_ID    TYPE      TABLE_NUM
------------------------------ ------------ --------- ---------
X $ KQFTA                      4294950912   TABLE     0
X $ KQFVI                      4294950913   TABLE     1
X $ KQFVT                      4294951149   TABLE     2
X $ KQFDT                      4294950914   TABLE     3
```

```
......
GV$WAITSTAT                    4294951371 VIEW        65537
V$WAITSTAT                     4294950915 VIEW        65537
GV$BH                          4294951405 VIEW        65537

已选择 2898 行。
SQL> col view_name for a10
SQL> col view_definition format a80
SQL> select * from v$fixed_view_definition where view_name = 'v$database';

VIEW_NAME      VIEW_DEFINITION                         CON_ID
-------------------------------------------------------------------------------
-------------------------------------------------------------
V$DATABASE
select   DBID, NAME, CREATED, RESETLOGS_CHANGE#, RESETLOGS_TIME, PRI/OR_RESETLOGS_CHANGE#,
PRI/OR_RESETLOGS_TIME, LOG_MODE, CHECKPOINT_CHANGE#, ARCHIVE_CHANGE#, CONTROLFILE_TYPE,
CONTROLFILE_CREATED,              CONTROLFILE_SEQUENCE#, CONTROLFILE_CHANGE#,
CONTROLFILE_TIME, OPEN_RESETLOGS,      VERSION_TIME,    OPEN_MODE, PROTECTION_MODE,
PROTECTION_LEVEL, REMOTE_ARCHIVE, ACTIVATION#, SWITCHOVER#, DATABASE_ROLE,
ARCHIVELOG_CHANGE#, ARCHIVELOG_COMPRESSION,      SWITCHOVER_STATUS, DATAGUARD_BROKER,
GUARD_STATUS, SUPPLEMENTAL_LOG_DATA_MIN, SUPPLEMENTAL_LOG_DATA_PK, SUPPLEMENTAL_LOG_DATA_UI,
FORCE_LOGGING,    PLATFORM_ID,     PLATFORM_NAME, RECOVERY_TARGET_INCARNATION#, LAST_OPEN
_INCARNATION#, CURRENT_SCN, FLASHBACK_ON, SUPPLEMENTAL_LOG_DATA_FK, SUPPLMENTAL_LOG_DATA_
ALL, DB_UNIQUE_NAME, STANDBY_BECAME_PRIMARY_SCN, FS_FAILOVER_MODE, FS_FAILOVER_STATUS, FS_
FAILOVER_CURRENT_TARGET, FS_FAILOVER_THRESHOLD, FS_FAILOVER_OBSERVER_PRESENT, FS_FAILOVER_
OBSERVER_HOST,
CONTROLFILE_CONVERTED, PRIMARY_DB_UNIQUE_NAME,      SUPPLEMENTAL_LOG_DATA_PL, MIN_
REQUIRED_CAPTURE_CHANGE#, CDB, CON_ID,    PENDING_ROLE_CHANGE_TASKS,    CON_DBID, FORCE_
FULL_DB_CACHING, SUPPLEMENTAL_LOG_DATA_SR
from GV$DATABASE where inst_id = USERENV('Instance')
```

可以使用 v$fixed_view_definition 视图查询到 v$视图和 gv$视图的定义。在$Oracle_HOME/RDBMS/ADMIN/CATALOG.SQL 中执行了 cdfixed.sql 脚本。在这个脚本中可以找到 gv_、v_$同义词的创建。

（1）gv$：全局视图，针对多个实例环境。
（2）v$：针对某个实例的视图。
（3）x$：gv$视图的数据来源，Oracle 数据库的内部表。
（4）gv_$：gv$的同义词。
（5）v_$：v$的同义词。

2. 常用的动态性能视图

（1）v$controlfile：显示控制文件列表。

（2）v$database：从控制文件中获取的数据库的信息。

（3）v$datafile：显示数据文件的信息。

（4）v$instance：显示当前实例的状态。

（5）v$parameter：显示内存中的参数信息。

（6）v$session：显示当前会话的信息。

（7）v$sga：显示 SGA 的信息。

（8）v$spparameter：显示 spfile 文件中的参数信息，如果 spfile 文件中没有被用来启动的实例，则视图中的所有 isspecified 列的值将会是 FALSE。

（9）v$tablespace：从控制文件中获取的表空间的信息。

（10）v$thread：从控制文件中获取的线程的信息。

（11）v$version：获取关键组件的版本信息。

4.3 项目案例

在日常使用 Oracle 数据库中经常用到数据字典与动态性能视图。下面以常见的数据字典与动态性能视图为例进行实验。

视频讲解

步骤 1　查询数据库名字、创建日期等。

```
SQL> col name format a30
SQL> select name, object_id, type, table_num from v$fixed_table;

NAME                         OBJECT_ID    TYPE    TABLE_NUM
---------------------------  -----------  ------  ---------
X $ KQFTA                    4294950912   TABLE   0
X $ KQFVI                    4294950913   TABLE   1
X $ KQFVT                    4294951149   TABLE   2
X $ KQFDT                    4294950914   TABLE   3
……
GV $ WAITSTAT                4294951371   VIEW    65537
V $ WAITSTAT                 4294950915   VIEW    65537
GV $ BH                      4294951405   VIEW    65537
已选择 2898 行.
```

步骤 2　查询数据库计算机的主机名、实例名和数据库系统版本。

```
SQL> select host_name,instance_name,version from v$instance;

HOST_NAME                    INSTANCE_NAME    VERSION
---------------------------  ---------------  ----------
LAPTOP - 8BM7LBRI            neworcl          19.0.0.0.0
```

步骤3　查询所安装 Oracle 数据库系统的版本方面的详细信息。

```
SQL> select * from v$version;

BANNER
BANNER_FULL
BANNER_LEGACY
CON_ID
--------------------------------------------------------------------------------
Oracle Database 19c Enterprise Edition Release 19.0.0.0.0 - Production
Oracle Database 19c Enterprise Edition Release 19.0.0.0.0 - Production Version 19.3.0.0.0
Oracle Database 19c Enterprise Edition Release 19.0.0.0.0 - Production          0
```

步骤4　查询控制文件等信息。

```
SQL> select name from v$controlfile;

NAME
--------------------------------------------------
D:\ORACLE19C\ORADATA\NEWORCL\CONTROL01.CTL
D:\ORACLE19C\ORADATA\NEWORCL\CONTROL02.CTL
D:\BACKUP\CONTROL03.CTL
```

步骤5　查询日志文件等相关信息。

```
SQL> select group#,members,bytes,status,archived from v$log;

    GROUP#    MEMBERS    BYTES      STATUS        ARC
 ----------- ---------- ---------- ------------- -----
        1          1    209715200  INACTIVE      YES
        2          2    209715200  CURRENT       NO
        3          3     15728640  ACTIVE        YES

SQL> select group#,member from v$logfile;

    GROUP# MEMBER
 --------- -----------------------------------------
        2  D:\ORACLE19C\ORADATA\NEWORCL\REDO02-2.LOG
        2  D:\ORACLE19C\ORADATA\NEWORCL\REDO02.LOG
        1  D:\ORACLE19C\ORADATA\NEWORCL\REDO01.LOG
        3  D:\ORACLE19C\ORADATA\NEWORCL\REDO03-1.LOG
        3  D:\ORACLE19C\ORADATA\NEWORCL\REDO03-2.LOG
        3  D:\ORACLE19C\ORADATA\NEWORCL\REDO03-3.LOG

已选择 6 行.
```

步骤 6　查询表空间相关信息。

```
SQL> select tablespace_name,block_size,status,contents,logging from dba_tablespaces;

TABLESPACE   BLOCK_SIZE   STATUS    CONTENTS           LOGGING
----------   ----------   -------   ----------------   ---------
SYSTEM             8192   ONLINE    PERMANENT          LOGGING
SYSAUX             8192   ONLINE    PERMANENT          LOGGING
UNDOTBS1           8192   ONLINE    UNDO               LOGGING
TEMP               8192   ONLINE    TEMPORARY          NOLOGGING
USERS              8192   ONLINE    PERMANENT          LOGGING
```

步骤 7　查询数据文件相关信息。

```
SQL> select file_id,file_name,tablespace_name,bytes/1024/1024 MB from dba_data_files;

FILE_ID FILE_NAME                                      TABLESPACE         MB
------- ---------------------------------------------  ----------------   -----
      1 D:\ORACLE19C\ORADATA\NEWORCL\SYSTEM01.DBF      SYSTEM             910
      3 D:\ORACLE19C\ORADATA\NEWORCL\SYSAUX01.DBF      SYSAUX             720
      7 D:\ORACLE19C\ORADATA\NEWORCL\USERS01.DBF       USERS                5
      4 D:\ORACLE19C\ORADATA\NEWORCL\UNDOTBS01.DBF     UNDOTBS1            65
```

步骤 8　查询数据库到底有多少用户及其创建时间。

```
SQL> col username for a50
SQL> select username,created from dba_users;

USERNAME                                             CREATED
-------------------------------------------------    --------------
SYS                                                  30-5月 -19
SYSTEM                                               30-5月 -19
XS$NULL                                              30-5月 -19
OJVMSYS                                              30-5月 -19
LBACSYS                                              30-5月 -19
OUTLN                                                30-5月 -19
SYS$UMF                                              30-5月 -19
DBSNMP                                               30-5月 -19
APPQOSSYS                                            30-5月 -19
DBSFWUSER                                            30-5月 -19
GGSYS                                                30-5月 -19
HR                                                   25-10月 -19
.......
已选择 37 行.
```

第 2 部分

数据库管理

第 5 章　Oracle 数据库常用工具

第 6 章　Oracle 空间管理

第 7 章　Oracle 网络配置管理

第 8 章　Oracle 监控管理

第 9 章　数据库的归档模式管理

第5章

Oracle数据库常用工具

PPT 视频讲解

Oracle 数据库提供了一套全面的、适用于开发者以及管理人员使用的多种工具。下面从开发人员角度和管理人员角度来讲解 Oracle 的常用工具。

5.1 数据库开发工具

应用开发工具和商务智能工具可支持任何开发方法、技术平台和操作系统。开发人员使用这些工具构建复杂的应用,从而为驱动解决方案的数据提供更高的可用性、可见性和可管理性。其中,SQL Plus 是 Oracle 数据库系统默认安装自带的一个客户端工具。可以在程序里直接打开,也可以通过命令行中输入 sqlplus 命令来启动该工具。

Oracle 数据库还为开发者提供了以下 5 个常用的通用工具。

1. SQL Developer

SQL Developer 是 SQL Plus 的一个图形化的版本,支持 SQL 和 PL/SQL 开发。可以使用标准的数据库身份验证并连接到任何 Oracle 数据库模式。

SQL Developer 的功能可以归纳如下:

(1) 浏览、创建、编辑和删除模式对象;

(2) 运行 SQL 语句;

(3) 编辑调试 PL/SQL 程序单元;

(4) 操作和导出数据;

(5) 创建和显示报表。

在官方网站中即可自行下载 SQL Developer,最新的版本为 SQL Developer 19.2.1。下载地址为:https://www.Oracle.com/tools/downloads/sqldev-v192-downloads.html。

Oracle SQL Developer 软件如图 5-1 所示。

2. Oracle Application Express(APEX)

APEX(Oracle 数据库应用程序快速开发工具)是针对 Oracle 数据库的 Web 应用程序

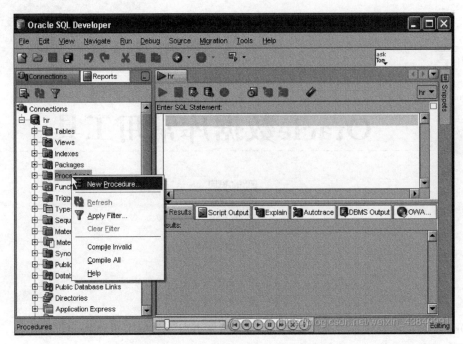

图 5-1　Oracle SQL Developer 软件

开发工具。该工具使用一些内置的功能如用户界面主题、导航控件、表单处理程序和灵活的报表来加快应用程序的开发。

APEX 是一个构建 Web 应用程序的工具,而且应用程序开发环境也基于 Web,非常方便。只需在 apex.Oracle.com 注册一个账户即可使用 APEX。开发、部署或者运行 APEX 应用程序无须客户端软件,当然也可以在本机计算机安装一套完整的 APEX 操作环境。

APEX 软件界面如图 5-2 所示。

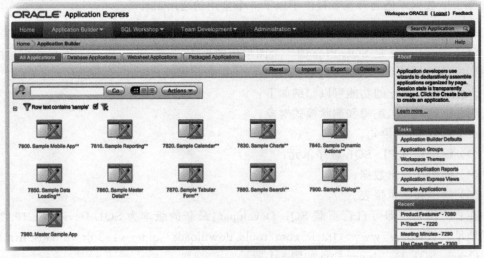

图 5-2　APEX 软件界面

3. Oracle JDevelper

Oracle JDevelper 是一个集成开发环境,它提供了对建模、开发、调试、优化和部署 Java 的应用程序,也是 Web 服务器的端到端支持。Oracle JDevelper 支持完整的软件开发周期,具有建模、编码、测试、调试、分析、优化和部署应用程序等集成的功能。

4. Oracle JPublisher

Oracle JPublisher 是一个简单方便的工具,用于创建访问数据库表的 Java 程序。

5. Oracle Developer Tools for Visual Studio.net

Oracle Developer Tools for Visual Studio.net 是一组与 Visual Studio.net 集成在一起的应用程序开发工具。

这些工具提供图形用户界面来访问 Oracle 数据库功能,使用户能够执行广泛的应用程序开发任务,提高开发效率和易用性。

5.2 数据库开发人员的主题

作为数据库开发人员,工作主题可以归纳如下:
(1) 应用程序设计和优化原则;
(2) 客户端数据库编程;
(3) 全球化支持;
(4) 非结构化数据。

5.3 数据库管理工具

作为数据库管理员,其常用的工具可以归纳如下:

1. Oracle 企业管理器

OEM 企业管理器(Oracle Enterprise Manager)是一个提供数据库环境集中化管理的系统管理工具。其将图形控制台、Oracle 管理服务器、Oracle 智能代理、公共服务和管理工具结合在一起,为 Oracle 数据库产品提供一个综合的系统管理平台。

OEM 的具体功能如下:
(1) 为数据库管理员提供的一个集中的系统管理工具;
(2) 一个用来管理、诊断和调试(调优)多个数据库的工具;
(3) 一个用来管理来自多个地点的多个网络节点和服务的工具;
(4) 方便不同的数据库管理员之间共享工作;
(5) 提供一些管理并行服务器和分布式数据库的工具。

若要获取 OEM 控制台的端口号,可在 $ORACLE_HOME\INSTALL 目录下的 portlist.ini 正文文件中查找。若要获取 OEM 的地址,则可通过查询同目录下的 readme.txt 文件即可。

OEM 界面如图 5-3 所示。

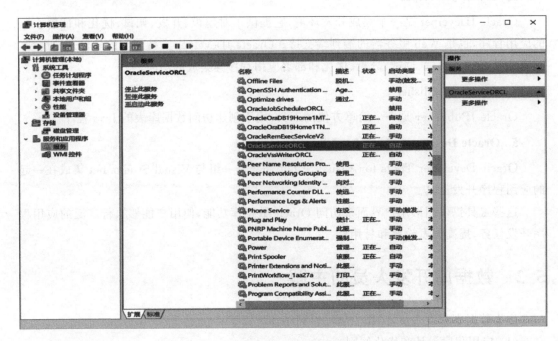

图 5-3 OEM 界面

2. SQL Plus

SQL Plus 是一个包含在每个 Oracle 数据库安装中的交互式和批处理查询工具,当连接到数据库时,可作为客户端的命令行用户接口。

SQL Plus 有其自己的命令和环境。它可输入 SQL、PL/SQL、SQL Plus 和操作系统命令来执行任务。

SQL Plus 工具如图 5-4 所示。

图 5-4 SQL Plus 工具

常用的 SQL Plus 命令如下:
(1) SQL > show all:查看所有 68 个系统变量值。
(2) SQL > show user:显示当前连接的用户。
(3) SQL > show error:显示错误。

(4) SQL＞set heading off：禁止输出列标题，默认值为 ON。

(5) SQL＞set feedback off：禁止显示最后一行的计数反馈信息，默认值为"对 6 个或更多的记录，回送 ON"。

(6) SQL＞set timing on：默认为 OFF，设置查询耗时，可用来估计 SQL 语句的执行时间，测试性能。

(7) SQL＞set sqlprompt "SQL＞"：设置默认提示符，默认值就是"SQL＞"。

(8) SQL＞set linesize 1000：设置屏幕显示行宽为 1000(未设置时，默认为 100)。

(9) SQL＞set autocommit ON：设置是否自动提交，默认为 OFF。

(10) SQL＞set pause on：默认为 OFF，设置暂停，会使屏幕显示停止，在单击 Enter 键时，会显示下一页。

3．数据库安装和配置工具

Oracle 提供了一些工具来简化安装和配置 Oracle 数据库软件的任务，这些工具如下：

(1) OUI(Oracle Universal Installer，Oracle 通用安装程序)

OUI 是一个图形用户界面实用程序，可查看、安装、升级、卸载或删除软件组件和创建数据库。

(2) DBCA(Database Configuration Assistant，数据库配置助手)

DBCA 提供了一个图形界面和引导工作流以创建和配置数据库，此工具能够从 Oracle 提供的模板创建数据库，或者创建自己的数据库和模板。

4．Oracle 网络配置和管理工具

Oracle 网络服务提供企业范围的分布式异构计算机环境中的连接解决方案。Oracle 网络配置和管理工具是 Oracle 网络服务的一个组件，可将一个网络会话从客户端应用程序连接到服务器。

配置和管理 Oracle 网络服务的工具如下所述。

(1) Oracle Net manager(Oracle 网络管理器)。

(2) Oracle Net Configration Assistant(Oracle 网络配置助理)。

5．数据移动和分析工具

Oracle 数据库有以下 4 个实用程序来辅助数据的移动和分析。

(1) SQL＊Loader(SQL 加载器)：将数据从称为数据文件的外部文件加载到数据库表中，它有一个强大的数据分析引擎，对数据文件中的数据格式几乎没有什么限制。

(2) Oracle Data Pump Export and Import(Oracle 数据库导入和导出)：能够将数据和元数据，从一个数据库快速地移动到另一个数据库。

(3) Oracle Log Miner(Oracle 日志挖掘器)：通过 SQL 接口查询重做日志文件，查明、检查并分析系统行为和错误。

(4) ADR Command Interpreter(ADR 命令解释器)：是一个命令行实用程序，可以调查问题、查看健康检查报告、将首次故障针对数据打包并上传到 Oracle。

5.4 数据库管理人员的主题

数据库管理人员的工作重心不同于数据库开发人员,其工作主题可以归纳如下:
(1) 备份和恢复;
(2) 内存管理;
(3) 资源管理与任务调度;
(4) 性能诊断和调优。

5.5 项目案例

在安装数据库软件的时候默认创建了实例数据库,但是在安装完毕后还可以再次单独创建实例数据库。下面借助 DBCA 图形化建库工具来感受一下创建数据库的简便步骤。

步骤 1　选择"开始"→Oracle OraDB19Home1→Database Configuration Assistant 菜单,如图 5-5 所示。

图 5-5　Database Configuration Assistant 菜单

步骤 2　在弹出的"选择数据库操作"对话框中选择"数据库操作"→"创建数据库"选项,单击"下一步"按钮,如图 5-6 所示。

步骤 3　为了灵活创建该数据库,在"选择数据库创建模式"对话框中选择"创建模式"→"高级配置"选项,单击"下一步"按钮,如图 5-7 所示。

步骤 4　在弹出的"选择数据库部署类型"对话框的"部署类型"选项卡中,在"数据库类型"下拉列表中选择"Oracle 单实例数据库"选项,单击"下一步"按钮,如图 5-8 所示。

步骤 5　在"指定数据库标识详细信息"对话框的"数据库标识"选项卡中对数据库进行实例配置。本书在安装数据库的环节创建的 neworcl 数据库为非容器数据库,所以在这个环节中,创建的 testorcl 数据库为容器数据库,如图 5-9 所示。

步骤 6　在"选择数据库存储选项"对话框的"存储选项"选项卡中对存储类型和位置进行设置,如图 5-10 所示。

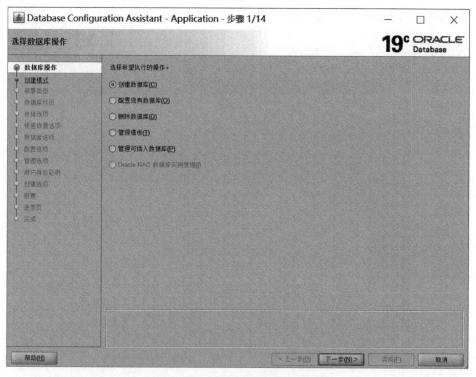

图 5-6 "选择数据库操作"对话框

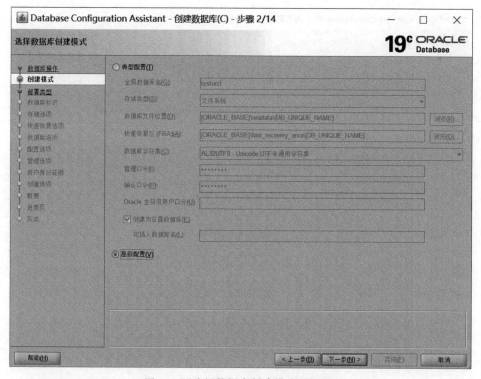

图 5-7 "选择数据库创建模式"对话框

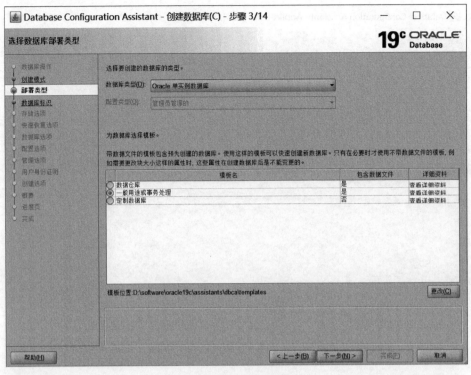

图 5-8 "部署类型"选项卡

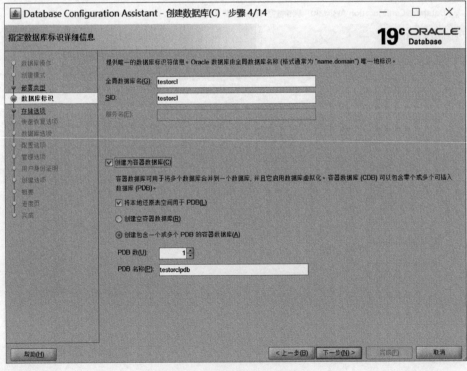

图 5-9 "数据库标识"选项卡

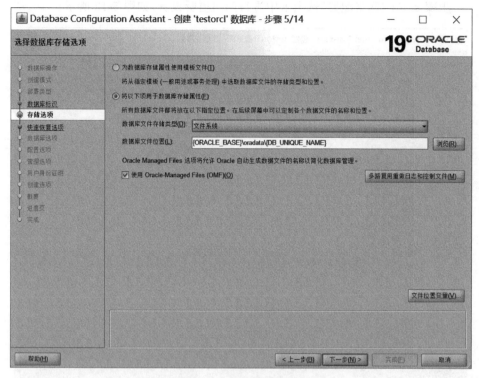

图 5-10 "存储选项"选项卡

步骤 7　在"快速恢复选项"选项卡中选择"启用归档"选项,如图 5-11 所示。

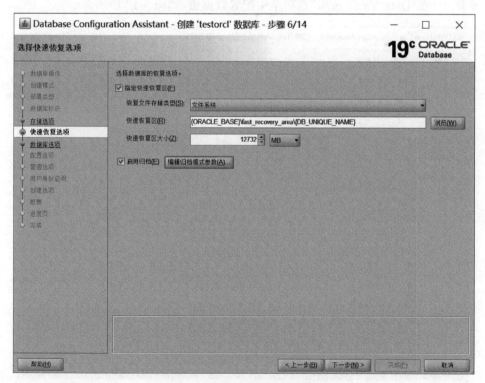

图 5-11 "快速恢复选项"选项卡

步骤 8　在"网络配置"选项卡中进行网络监听设置,如图 5-12 所示。

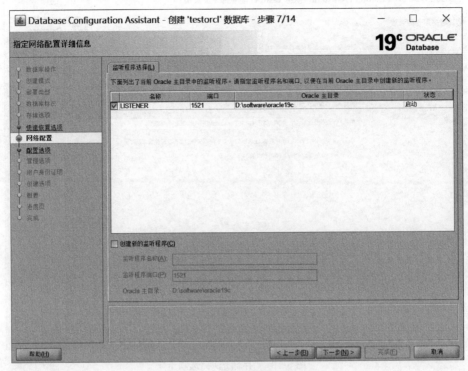

图 5-12　"网络配置"选项卡

步骤 9　在"Data Vault 选项"选项卡中配置数据值,如图 5-13 所示。

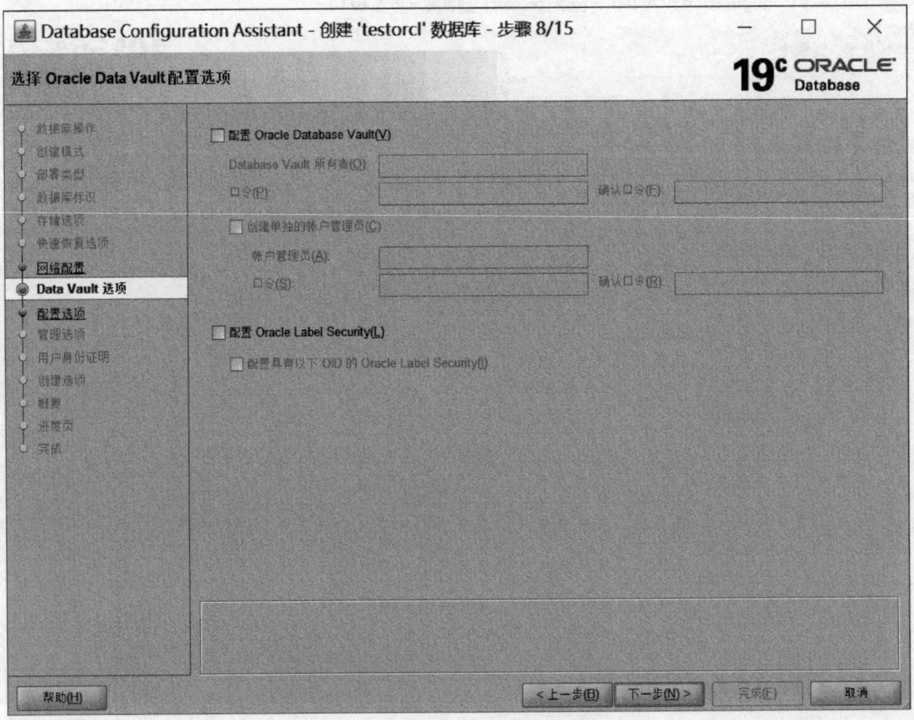

图 5-13　"Data Vault 选项"选项卡

步骤10 在"配置选项"选项卡可进行内存、调整大小、字符集、连接模式以及示例方案的设置,如图 5-14~图 5-18 所示。

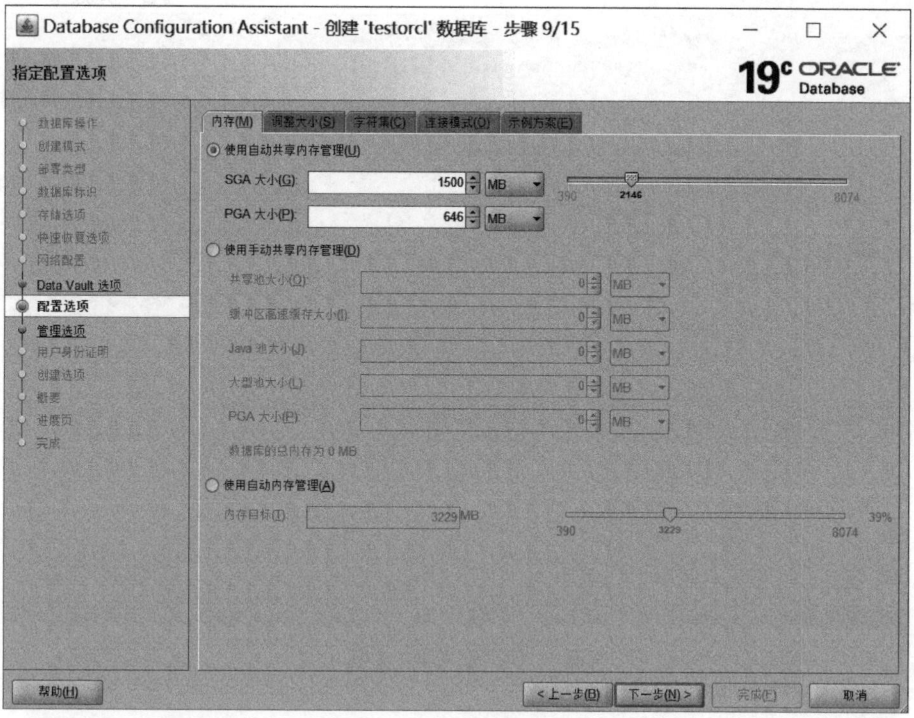

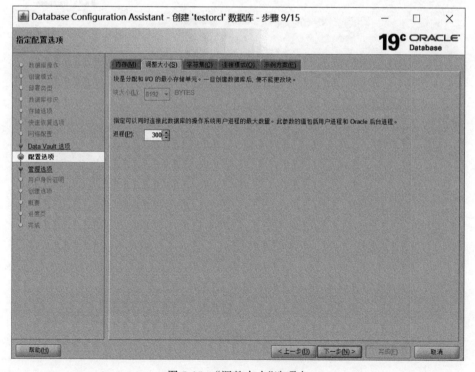

图 5-14 "内存"选项卡

图 5-15 "调整大小"选项卡

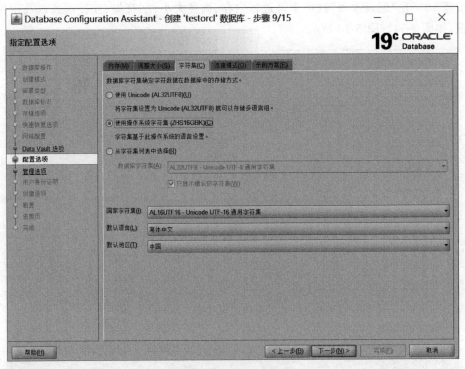

图 5-16 "字符集"选项卡

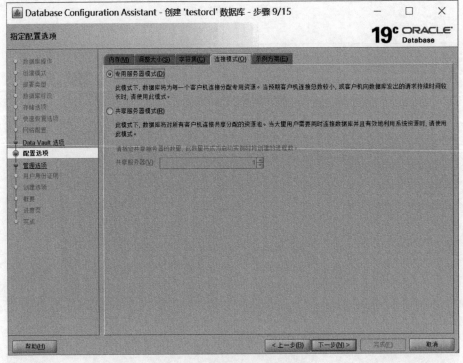

图 5-17 "连接模式"选项卡

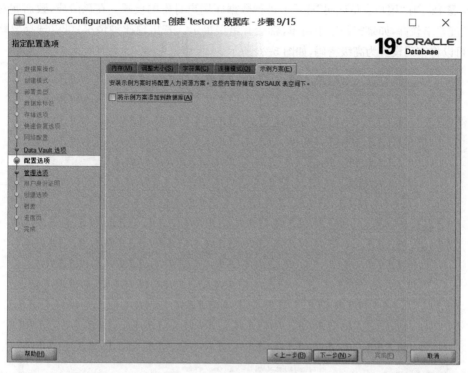

图 5-18 "示例方案"选项卡

步骤 11 在"管理选项"选项卡中进行数据库管理选项 EM 的设置,如图 5-19 所示。

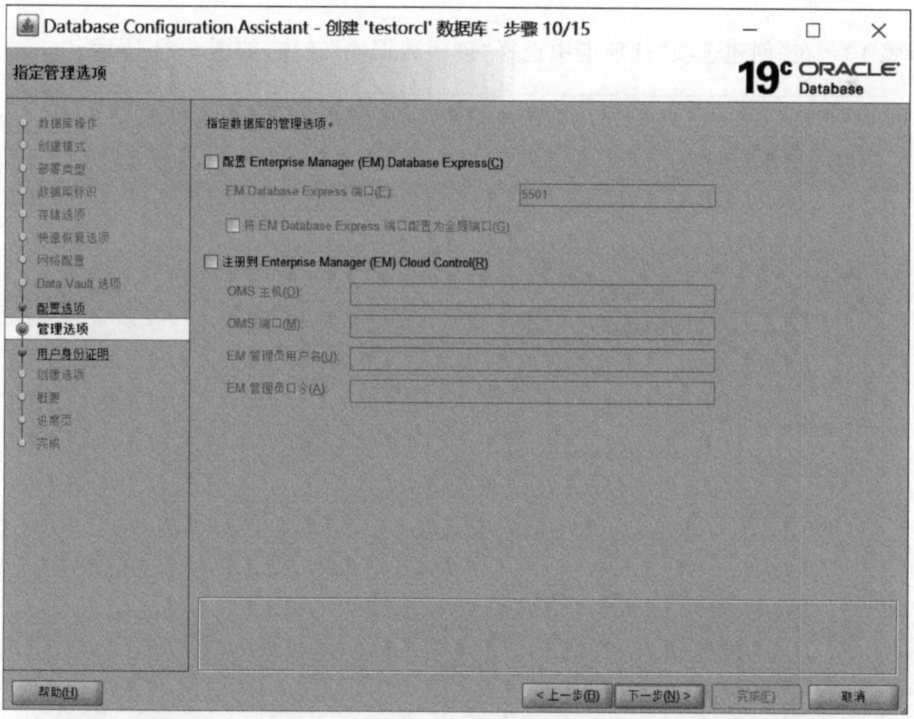

图 5-19 "管理选项"选项卡

步骤 12 在"用户身份证明"选项卡进行数据库用户身份的设置。在该阶段，数据库对密码的复杂度要求较高，如果密码过于简单，就会有错误提示。若出现此种情况，可以选择忽略该提示，也可以将密码设置为高级密码，如图 5-20 所示。

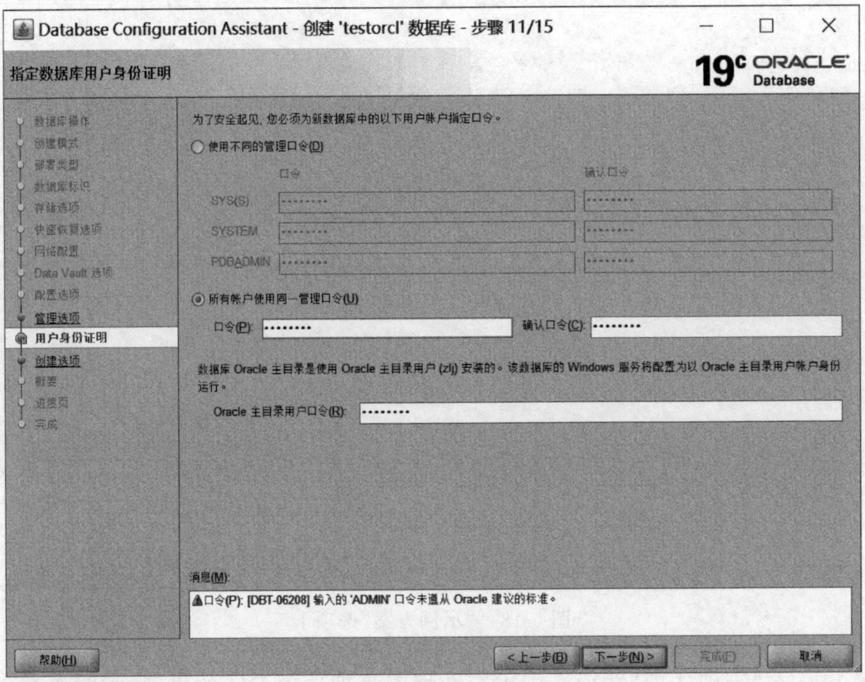

图 5-20 "用户身份证明"选项卡

步骤 13 在"创建选项"选项卡中选择"创建数据库"选项，如图 5-21 所示。

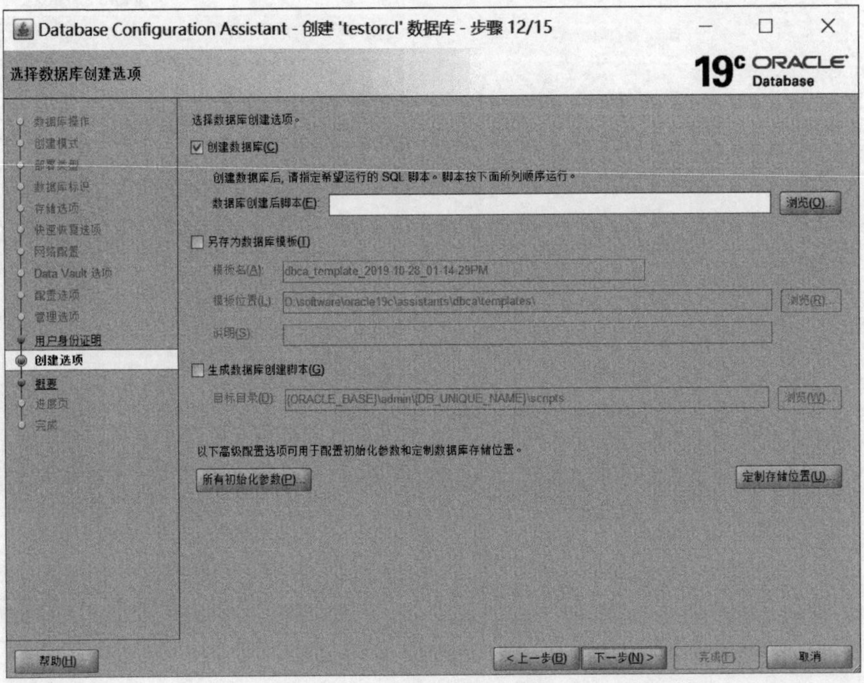

图 5-21 "创建选项"选项卡

步骤 14　在"概要"对话框中进行概要总结，如图 5-22 所示。

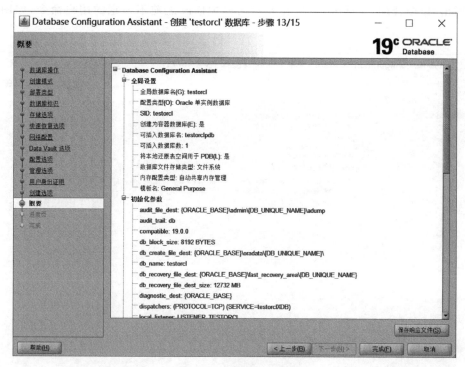

图 5-22　"概要"对话框

步骤 15　在"进度页"选项卡中显示安装进度，安装时间较长，需要耐心等待。最后会提示安装成功，如图 5-23 所示。

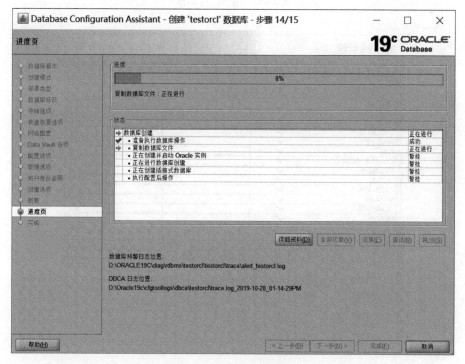

图 5-23　"进度页"对话框

步骤16 最后为创建完成阶段,如图5-24所示;在"完成"对话框中,单击"口令管理"按钮,弹出"口令管理"对话框,可以在此对话框中修改之前不满意的用户命令,如图5-25所示。

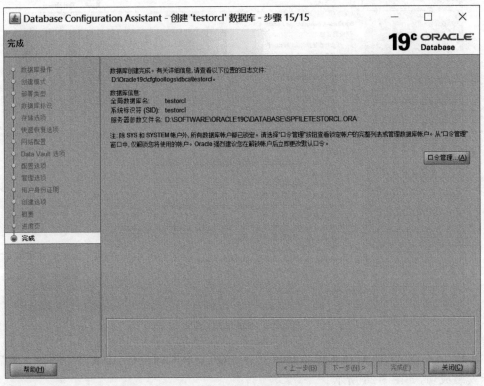

图5-24 "完成"对话框

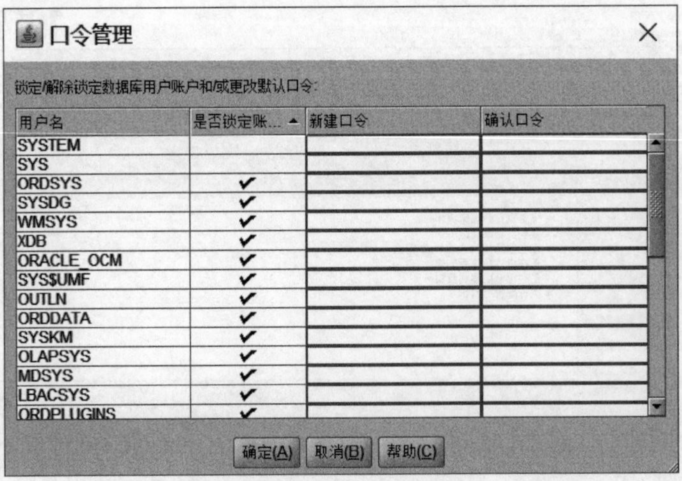

图5-25 "口令管理"对话框

第6章 Oracle空间管理

PPT 视频讲解

严格来说,数据库是由数据文件、控制文件、日志文件等组成的。在正常情况下,数据文件占用了绝大多数的操作系统空间。

本章首先通过以下脚本观察数据库空间的使用情况。

```
SQL>set serveroutput on;
SQL>declare
  2    V_b number;
  3    V_C number;
  4    V_d number;
  5    V_e number;
  6  begin
  7    dbms_output.put_line(chr(10));
  8    select sum(bytes)/1024/1024 into v_b from v$datafile;
  9    select sum(bytes)/1024/1024 into v_c from dba_free_space;
 10    v_d: = v_b - v_c;
 11    v_e: = trunc((v_b - v_c)/v_b * 100,2);
 12    dbms_output.put_line('本次Oracle检查数据:');
 13    dbms_output.put_line('数据库空间分配总规模:'||v_b||'(MB)');
 14    dbms_output.put_line('数据库空闲空间总规模:'||v_c||'(MB)');
 15    dbms_output.put_line('数据库已使用空间'||v_d||'(MB)');
 16    dbms_output.put_line('已使用总空间比率'||v_e||'%');
 17  end;
 18  /

本次Oracle检查数据:
数据库空间分配总规模:1700(MB)
```

```
数据库空闲空间总规模: 106.625(MB)
数据库已使用空间 1593.375(MB)
已使用总空间比率 93.72 %

PL/SQL 过程已成功完成。
```

空间管理涉及很多内容,比如最重要的表空间的管理等。很多DBA喜欢在建库之初就给数据库分配很多空闲空间,这种方法虽然可以避免数据库空间不足,但会导致数据库的空闲空间很大,而过大的空闲空间可能带来很多副作用,如增加数据库的管理成本、影响数据库的打开和关闭时间、增加数据库的迁移难度等。基于以上考虑,一般建议给数据库预留足够的空闲空间即可。一套运行良好的数据库是"用心"管理出来的,而不是空间预分配出来的。

6.1 数据库的创建与配置

数据库的创建可采用DBCA图形工具、静默方式和手动创建三种方式。

通常创建Oracle数据库都是通过DBCA(Database Configuration Assistant,数据库配置助手)工具完成的,DBCA工具可以通过两种方式完成建库任务,即图形界面方式和静默命令行方式。在本书的5.5节的"项目案例"中按步骤给大家演示了建库过程,关于静默方式将不再赘述。

手动创建数据库比使用DBCA创建数据库要麻烦,但是手动创建数据库比通过DBCA创建更加可控,也适用于图形界面无法启动的主机环境,手动创建数据库能比较清楚地了解创建数据库的来龙去脉。如果学好了手动创建数据库,就可以更好地理解Oracle数据库的体系结构。

手动创建数据库需要经过如下步骤,每一步都非常关键。

(1) 创建必要的相关目录。
(2) 创建初始化参数文件。
(3) 设置环境变量 oracle_sid。
(4) 创建实例。
(5) 创建口令文件。
(6) 启动数据库到 Nomount 状态。
(7) 执行建库脚本。
(8) 执行 catalog 脚本创建数据字典。
(9) 执行 catproc 创建 package 包。
(10) 执行 pupbld。
(11) 由初始化参数文件创建 spfile 文件。
(12) 执行 hr 脚本创建 hr 模式。

6.2 表空间的管理

Oracle表空间是Oracle数据对象和数据存储的容器,Oracle表空间经常和数据文件成对出现,一个表空间可以对应多个数据文件,而一个数据文件只能在一个表空间中。当创建表空间时,就会默认创建一个数据文件,同理,当创建数据文件时,必须指定一个表空间。

Oracle 的数据(表、索引等数据)存储在数据文件中,在表空间中的逻辑单位是段。例如,创建一个索引时,会在指定表空间下创建一个以索引名字命名的索引段,然后在索引段中创建一个或者多个区用来存储索引数据,一个区只能存在于一个数据文件中。一个区中可以分为多个块(Block)。块是 Oracle 数据库当中最小的空间分配单位。

一般地,一个文件在磁盘空间中存储都不是连续的,因此,表空间中的段是由不同数据文件中的区组成的。

表空间管理虽然较为复杂,但是数据库管理员却可以利用表空间完成很多工作,如下所述:

(1) 控制数据库数据的磁盘分配。
(2) 将确定的空间份额分配给数据库用户。
(3) 通过使单个表空间联机或脱机,控制数据的可用性。
(4) 执行部分数据库备份或恢复操作。
(5) 为提高性能,进行设备分配数据存储。

在成功地安装 Oracle 19c 后,默认存在 5 个表空间,执行以下指令可查看表空间。

```
SQL> select tablespace_name, initial_extent, extent_management, allocation_type from dba_tablespaces;

TABLESPACE    INITIAL_EXTENT    EXTENT_MAN    ALLOCATION
----------    --------------    ----------    ----------
SYSTEM                 65536    LOCAL         SYSTEM
SYSAUX                 65536    LOCAL         SYSTEM
UNDOTBS1               65536    LOCAL         SYSTEM
TEMP                 1048576    LOCAL         UNIFORM
USERS                  65536    LOCAL         SYSTEM
```

其中:

SYSTEM 表空间:存储 sys/system 用户表、存储过程、视图等数据库对象;

SYSAUX 表空间:安装示例的数据库表空间;

UNDOTBS1 表空间:用于存储撤销(用于回滚)的信息;

TEMP 表空间:临时表空间,用于存储 SQL 语句处理的表和索引信息;

USERS 表空间:存储数据库用户创建的数据库对象信息。

6.2.1 表空间的管理方式

Oracle 通过表空间为数据库提供使用空间。由于区是 Oracle 创建对象时的最小分配单元,所以表空间的管理实际上就是针对于区的管理。

Oracle 表空间的管理方式有以下两种。

1. 字典管理方式

使用数据字典管理存储空间的分配,当表空间分配新的区,或者回收已分配的区时,

Oracle 会对数据字典对应的表进行查询、更新。这种管理方式的缺点是使用单线程、速度慢,且会产生回退和重做信息。

2. 本地管理方式

表空间中区分配和区回收的管理信息都被存储在表空间的数据文件中,与数据字典无关。表空间为每个数据文件维护一个位图结构,用于记录表空间的区分配情况。由于当表空间分配新的区,或者回收已分配的区时,Oracle 会对文件中的位图进行更新,所以不会产生回滚和重做信息。

本地管理方式具有以下 4 个特点。

(1) 提高存储管理的速度和并发性。
(2) 不产生磁盘碎片。
(3) 不产生递归管理。
(4) 没有系统回滚段。

6.2.2 表空间管理

1. 创建表空间

在创建本地管理方式下的表空间时,应该确定表空间的名称、类型、对应的数据文件的名称和位置以及区的分配方式、段的管理方式。

创建表空间的语法如下:

```
create tablespace table_name
datafile'filename'
size n
[autoextend on next n1 maxsize m /off]
[permanent]
[extent management local/dictionary];
```

其中:

create tablespace:创建表空间的关键字;

table_name:创建后表空间的名字;

filename:指定数据文件的路径;

size n:指定数据文件的大小;

autoextend on next n1 maxsize m/off:表空间是否是自动扩展的,on 为自动扩展,off 为不扩展;当自动扩展时,next n1 表示自动扩展的大小,max size m 表示数据文件最大可扩展到 m 大小;

permanent:创建的表空间的类型,permanent 表示永久表空间,不填都是默认永久表空间;

extent management local/dictionary:表空间管理的方式,local 表示本地的管理模式,dictionary 表示数据字典管理模式,默认为本地管理方式。

下面为大家示例创建一个本地表空间——TEST,其代码如下:

```
SQL>create tablespace test
  2      datafile 'D:\ORACLE19C\ORADATA\NEWORCL\TEST01.DBF'
  3      size 25m autoextend on next 1280k maxsize 1000m,
  4              'D:\ORACLE19C\ORADATA\NEWORCL\TEST02.DBF'
  5      size 25m autoextend on next 1280k maxsize 1000m
  6      extent management local uniform size 64k;
```

表空间已创建.

2. 查看表空间

如何查看数据库有哪些表空间？如何查看表空间对应的数据文件？

下面以常见的数据字典视图和动态性能视图进行有关表空间相关信息的查询。

（1）包含数据库中所有表空间的描述信息查询：

```
SQL> select * from dba_tablespaces
```

（2）包含当前用户的表空间的描述信息查询：

```
SQL> select * from user_tablespaces
```

（3）包含从控制文件中获取的表空间名称和编号信息的查询：

```
SQL> select * from v$tablespace;
```

（4）包含数据文件以及所属的表空间的描述信息查询：

```
SQL> select * from dba_data_files
```

（5）包含临时数据文件以及所属的表空间的描述信息查询：

```
SQL> select * from dba_temp_files
```

（6）包含从控制文件中获取的数据文件的基本信息，包括它所属的表空间名称、编号等的查询：

```
SQL> select * from v$datafile
```

（7）包含所有临时数据文件的基本信息的查询：

```
SQL> select * from v$tempfile
```

输入上面的部分命令，即可观察到新建的表空间 TEST 的情况，如下所示：

```
SQL>select * from v$tablespace;

     TS#  NAME                            INC  BIG  FLA ENC     CON_ID
    ----- ------------------------------  ---  ---  ---------  --------
       1  SYSAUX                          YES  NO   YES               0
       0  SYSTEM                          YES  NO   YES               0
       2  UNDOTBS1                        YES  NO   YES               0
       4  USERS                           YES  NO   YES               0
       3  TEMP                            NO   NO   YES               0
       9  TEST                            YES  NO   YES               0

已选择 6 行.
```

此外,继续观察新建的数据文件的情况。

```
SQL>select file#,name,bytes from v$datafile;

    FILE#  NAME                                                    BYTES
    -----  ------------------------------------------------  -----------
        1  D:\ORACLE19C\ORADATA\NEWORCL\SYSTEM01.DBF           954204160
        3  D:\ORACLE19C\ORADATA\NEWORCL\SYSAUX01.DBF           754974720
        4  D:\ORACLE19C\ORADATA\NEWORCL\UNDOTBS01.DBF           68157440
        5  D:\ORACLE19C\ORADATA\NEWORCL\TEST01.DBF              26214400
        7  D:\ORACLE19C\ORADATA\NEWORCL\USERS01.DBF              5242880
        8  D:\ORACLE19C\ORADATA\NEWORCL\TEST02.DBF              26214400

已选择 6 行.
```

3. 调整表空间

(1) 重置表空间大小。如果是本地管理的表空间,那么表空间的存储设置是不能改变的。但是用户可以用以下两个方法增加表空间的大小。

方法 1:使用 alter tablespace 语句增加数据文件,代码如下:

```
alter database datafile add/drop datafile 'filename'
```

方法 2:改变数据文件的大小,代码如下:

```
alter database datafile'filename'[autoextend on next n1];
——重新设置数据文件自动扩展参数
alter database datafile'filename' resize [n1];
——重置数据文件大小
```

(2) 移动表空间。如果某个磁盘 I/O 过于繁忙,就会影响数据库整体的性能。此时可

以将一个或多个数据文件移动到其他的磁盘上,以平衡 I/O。移动表空间的代码如下:

```
alter database  tablespace_name rename'old filename' to 'new filename';
```

(3)删除表空间。当一个表空间不再需要时,就可使用命令将其删除。但是有两种表空间不可以随意删除,第一种是系统表空间,第二种是有活动段的表空间。

删除表空间的语法如下:

```
drop tablespace tablespace_name
[including contents[and datafiles][casscade constraints]];
```

其中:

drop tablespace:删除表空间的关键字;

tablespace_name:表示表空间名字;

[including contents]:在删除表空间的时候把表空间中的数据文件一并删除;

[cascade constraints]:在删除表空间的时候把表空间的完整性也一并删除。比如表的外键和触发器等就是表的完整性约束。

利用删除表空间命令将刚才临时创建的表空间 TEST 删掉,同时级联删除表空间的所有内容以及数据文件,其代码如下:

```
SQL>drop tablespace test including contents and datafiles;
```

表空间已删除.

4. 维护表空间

(1)表空间脱机与联机的代码如下:

```
alter tablespace  tablespace_name offline immediate;
alter tablespace  tablespace_name online;
```

(2)表空间只读与读写的代码如下:

```
alter tablespace  tablespace_name ready only;
alter tablespace  tablespace_name read write;
```

(3)表空间自动扩展开启与关闭的代码如下:

```
alter database datafile'filename' autoextend on;
alter database datafile'filename' autoextend off;
```

6.2.3 还原表空间

由于还原(Undo)一词在以前的版本中被称为回滚(Roll Back),所以还原表空间是用来自动管理还原(回滚)数据的。

1. 查看还原表空间信息

使用 show 命令可以查询 undo 参数的设置情况,结果如下所示:

```
SQL> show parameter undo

NAME                                 TYPE        VALUE
------------------------------------ ----------- ----------
temp_undo_enabled                    boolean     FALSE
undo_management                      string      AUTO
undo_retention                       integer     900
undo_tablespace                      string      UNDOTBS1
```

观察结果需要了解下列参数的具体含义。

(1) temp_undo_enabled:在 Oracle 12c R1 版本之前,临时表生成的 undo 记录是存储在表空间 UNDO 里的,通用表和持久表的 undo 记录也是类似的。而在 Oracle 12c R1 版本后,临时 undo 功能中,临时 undo 记录可以存储在一个临时表中,而无须再存储在表空间 UNDO 内。temp_undo_enabled 默认值是 FALSE,因为这样不占用表空间,只要 commit 就清空。

(2) undo_management:是确定表空间 UNDO 的管理方式。如果该参数值设置为"AUTO",就表示系统使用自动 undo 管理;如果设置为"MANUAL",就表示使用手动 undo 管理,并以回滚段方式启动数据库。

(3) undo_retention:决定 undo 数据的维持时间,即用户事务结束后,undo 的保留时间,默认值为 900s。

(4) undo_tablespace:使用自动 undo 管理时,系统默认表空间 UNDO 名为 undotbs。

若要启用临时 undo 功能,则语法为:

```
alter system set temp_undo_enable = true;
```

2. 创建还原表空间

如果数据库中没有创建还原表空间,那么实例将使用表空间 SYSTEM 管理回滚段。若要手动创建还原表空间,可以采用以下两种方法创建还原表空间。

(1) 通过在 create database 命令中加入一个子句,在创建数据库时建立还原表空间。具体语法如下:

```
create database database_name
...
undo tablespace tablespace_name
datafile 'filename' size n1
autoextend on
```

（2）在创建数据库之后，使用 create undo tablespace 命令。具体语法如下：

```
create undo tablespace tablespace_name
datafile'filename'
size n;
```

3. 修改还原表空间

可以使用 alter tablespace 命令修改还原表空间。为还原表空间增加额外数据文件的语法如下：

```
alter tablespace tablespace_name
add datafile'filename'
size n1
autoextend on;
```

4. 激活还原表空间

在 Oracle 中，允许创建多个还原表空间，但是同一时间只能激活一个还原表空间。还原表空间之间的切换语法如下：

```
alter system set undo_tablespace = tablespace_name;
```

5. 删除还原表空间

删除还原表空间的语法如下：

```
drop tablespace tablespace_name;
```

在删除还原表空间的时候，需要注意以下事项：
（1）只有在一个还原表空间当前没有被使用时，才可以将它删除。
（2）删除一个活动的还原表空间的方法：
① 切换到一个新的还原表空间。
② 等所有当前的事物完成后再删除该还原表空间。

6. 还原信息相关的动态视图数据字典

通过查询视图可获取有关还原段的信息：dba_rollback_segs。与还原表空间相关的动态性能视图有：v$rollname、v$rollstat、v$undostat、v$session 和 v$transaction。

查看或查询临时 undo 数据相关统计信息的字典视图有：v$tempundostat、dba_hist_undostat 和 v$undostat。

6.2.4 临时表空间

通过 Oracle 表空间的讲解了解到表空间是 Oracle 数据库存储数据和对象的逻辑容器，而 Oracle 临时表空间主要是存储数据库的排序操作、临时表、中间排序结果等临时对象的。

例如，进行大数量级的排序操作时，当数据库内存不够时，就会写入临时表空间，当操作完成后，临时表空间就会自动清空释放。Oracle经常使用临时表空间的操作有创建索引（create index），分组查询（group by），排序时（order by），集合运算时（union，minus，intersect），多表连接查询时，当数据库内存不足时。如果在创建数据库时没有使用temporary tablespace子句，那么默认使用表空间SYSTEM作为排序区，但这样会降低数据库系统的效率。

1. 创建临时表空间

临时表空间既可以在创建数据库时创建，也可以在数据库创建执行后单独创建。

Oracle创建临时表空间的语法结构与创建持久化表空间一样，只是多了一个关键字——temporary。

创建临时表空间语法如下：

```
create temporary tablespace tempname
tempfile 'filename'
size m;
```

其中：

create temporary tablespace：创建临时表空间；

tempname：创建临时表空间的名字；

filename：指定临时表空间数据文件的位置；

size m：临时表空间的大小。

下面代码为临时表空间TEMP1的创建过程。

```
SQL>create temporary tablespace temp1
  2   tempfile 'D:\ORACLE19C\ORADATA\NEWORCL\TEMP1.DBF'
  3   size 50m;

表空间已创建.
```

2. 查询临时表空间

创建好临时表空间TEMP1，可以通过数据字典dba_temp_files进行查询临时表空间的信息，查询代码如下：

```
SQL>select t.tablespace_name, --表空间名
  2    t.file_name, --文件名
  3    t.autoextensible, --是否自动扩展
  4    t.bytes / 1024 / 1024 as tsize, --表空间初始大小
  5    t.maxbytes / 1024 / 1024 msize, --表空间最大扩展到多少
  6    b.contents, --表空间类型
  7    b.extent_management --表空间管理模式
  8    from dba_temp_files t, dba_tablespaces b
  9    where t.tablespace_name = b.tablespace_name;
```

```
TABLESPACE  FILE_NAME                           AUT TSIZE  MSIZE      CONTENTS  EXTENT_MAN
----------  ----------------------------------  --- -----  ---------  --------- ----------
TEMP        D:\ORACLE19C\ORADATA\NEWORCL\TEMP01.DBF  YES  317   32767.9844  TEMPORARY  LOCAL
TEMP1       D:\ORACLE19C\ORADATA\NEWORCL\TEMP1.DBF   NO   50    0           TEMPORARY  LOCAL
```

3. 修改临时表空间

如果没有显示创建临时表空间，Oracle 数据库在安装完成后就会创建一个默认的临时表空间 TEMP，其对应的文件为 temp01.dbf。在数据库创建后，若要修改默认的临时表空间，则需要使用 alter database 命令，具体代码如下：

```
SQL>alter database default temporary tablespace temp1;

数据库已更改。
```

4. 查看用户所用临时表空间的情况

在查看调整默认临时表空间后，用户所用临时表空间的情况的代码如下：

```
SQL>select username, temporary_tablespace from dba_users;

USERNAME                                  TEMPORARY_TABLESPACE
---------------------------------------   ------------------------
SYS                                       TEMP1
SYSTEM                                    TEMP1
HR                                        TEMP1
CTXSYS                                    TEMP1
DVSYS                                     TEMP1
DVF                                       TEMP1
........
已选择 37 行。
```

5. 删除临时表空间

如果要删除的临时表空间是用户的默认临时表空间，建议进行更改后再删除；否则可直接删除。具体删除临时表空间的执行代码如下：

```
SQL>alter database default temporary tablespace temp;

数据库已更改。

SQL>drop tablespace temp1 including contents and datafiles;
```

表空间已删除.

```
SQL>select tablespace_name from dba_tablespaces;

TABLESPACE
--------
SYSTEM
SYSAUX
UNDOTBS1
TEMP
USERS
```

6.2.5 用户与表空间

在学习完上述的表空间的管理后发现,表空间 SYSTEM 的空闲空间太小,表空间 USER 的空闲空间较大。这是为什么呢？这是因为如果在用户使用时,如果不为其分配表空间,那么系统就会按照默认的表空间进行分配。

6.2.6 表空间监控与注意事项

一般情况下,建议表空间的使用率不要超过 85%。当然还原表空间和临时表空间除外,如果超过了 85%,就须进行数据文件的扩展等操作。

网络上有很多与表空间相关信息的监控脚本,只要灵活地使用动态性能视图,就可以编写出适合的监控脚本。以下是一个监控表空间的使用率和其他属性的脚本,仅供读者参考。

```
SQL>select d.tablespace_name,
  2   space || 'm' "sum_space(m)",
  3   space - nvl(free_space, 0) || 'm' "used_space(m)",
  4   round((1 - nvl(free_space, 0) / space) * 100, 2) || '%'
  5   "used_rate(%)",
  6   free_space || 'm' "free_space(m)"
  7   from (select tablespace_name,
  8   round(sum(bytes) / (1024 * 1024), 2) space,
  9   sum(blocks) blocks
 10   from dba_data_files
 11   group by tablespace_name) d,
 12   (select tablespace_name,
 13   round(sum(bytes) / (1024 * 1024), 2) free_space
 14   from dba_free_space
 15   group by tablespace_name) f
 16   where d.tablespace_name = f.tablespace_name(+)
 17   union all
```

```
18    select d.tablespace_name,
19    space || 'm' "sum_space(m)",
20    used_space || 'm' "used_space(m)",
21    round (nvl (used_space, 0) / space * 100, 2) || '%' "used_rate(%)",
22    nvl (free_space, 0) || 'm' "free_space(m)"
23    from (select tablespace_name,
24    round (sum (bytes) / (1024 * 1024), 2) space,
25    sum (blocks) blocks
26    from dba_temp_files
27    group by tablespace_name) d,
28    (   select tablespace_name,
29    round (sum (bytes_used) / (1024 * 1024), 2) used_space,
30    round (sum (bytes_free) / (1024 * 1024), 2) free_space
31    from v$temp_space_header
32    group by tablespace_name) f
33    where d.tablespace_name = f.tablespace_name( + )
34    order by 1;

TABLESPACE_NAME   sum_space(m)   used_space(m)   used_rate(%)   free_space(m)
---------------   ------------   -------------   ------------   -------------
SYSAUX            720m           669.37m         92.97%         50.63m
SYSTEM            910m           902.75m         99.2%          7.25m
TEMP              317m           317m            100%           0m
UNDOTBS1          65m            28.62m          44.03%         36.38m
UNIFORMTB         50m            5m              10%            45m
USERS             5m             2.69m           53.8%          2.31m

已选择 6 行.
```

6.3 段的管理

段是 Oracle 中最重要的逻辑结构，其中表段可以用来存储，其他很多对象和结构都是围绕着表段而产生的，如索引段、回退段和临时段等。索引段是为了使表段中的数据检索更高效；回退段是为了记录标段数据的事务前镜像；临时段是为了对表段中排序操作存储中间数据。

下面以一个数据库表的创建展开段管理中存储参数的设置和优先级的相关操作。

```
SQL> create table test(
id number,
name varchar2(20)
)
tablespace uniformtb
storage(initial 1m next 512k minextents 5 pctincrease 0);
```

存储子句在段一级的说明，用于控制区在段中的分配，其分配原则如下所述。

(1) 任何在段一级说明的存储参数将覆盖在表空间一级所对应的选项设置,但是表空间级的参数 minimum extent 或 uniform size 除外。

(2) 当存储参数在段一级没有显式的定义时,它们默认为表空间一级所定义的参数值。

(3) 当存储参数在表空间一级没有显式的定义时,Oracle 数据库系统的默认参数值将被使用。

(4) 如果在表空间一级已经定义了 minimum extent 的大小,它将应用于该表空间中将来所有段的区的分配。

(5) 区某些存储参数不能在表空间一级定义,这些存储参数只能在段一级说明。

(6) 区如果对存储参数进行了修改,新的存储参数只适用于还没有分配的区。

存储子句的优先级如图 6-1 所示。

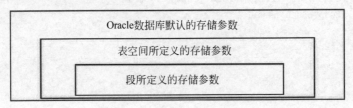

图 6-1 存储子句的优先级

6.4 区的管理

区是表空间中由某个段所使用的一块磁盘空间。在下列情况下,Oracle 会分配一个区:

(1) 区当一个段被创建(created)时。

(2) 区当一个段被扩展(extended)时。

(3) 区当一个段被改变(altered)时。

在下列情况下,Oracle 会回收一个区段:

(1) 区当一个段被删除(dropped)时。

(2) 区当一个段被改变(altered)时。

(3) 区当一个段被截断(truncated)时。

区可以分为已用区和空闲区,如图 6-2 所示。

6.5 块的管理

块是 Oracle 的最小存储单元,它由一个或多个操作系统块组成,它的大小在表空间创建时设置,db_block_size 为默认设置 Oracle 块大小的参数。

1. 数据块的大小

执行下面语句代码就可以查询默认块的大小。

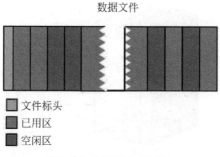

图 6-2　已用区与空闲区

```
SQL>show parameter db_block_size

NAME                                 TYPE        VALUE
------------------------------------ ----------- ------------------------------
db_block_size                        integer     8192
```

如果要在一个数据库中使用多种块,就必须定义 db_cache_size 和至少一个 db_nk_cache_size 参数,代码如下:

```
SQL>alter system set db_16k_cache_size = 1m;

系统已更改.
SQL>show parameter db_16k_cache_size

NAME                                 TYPE        VALUE
------------------------------------ ----------- ------------------------------
db_16k_cache_size                    big integer 32M
```

从运行结果中可发现,虽然用户指定了 1MB 的空间,但是系统为其分配了至少 32MB 的空间。

由上面创建的 16KB 的块可以进行表空间的创建。具体代码如下:

```
SQL>create tablespace DATA1 datafile 'D:\ORACLE19C\ORADATA\NEWORCL\DATA01.DBF' size 25m
uniform size 4m   blocksize 16k;

表空间已创建.
```

2. 数据块的结构

数据块的结构如图 6-3 所示。

其中,块的结构由以下 3 部分组成。

(1) 块头(Hearder):块头不仅记录一些控制信息,以帮助 Oracle 定位这个块,还记录块与块之间的串联信息。

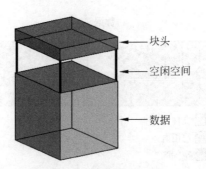

图 6-3 数据块的结构

(2) 空闲空间(Free Space)：处于空闲状态的空间。

(3) 数据(Data)：已经写入数据的空间。数据存放数据的方式是自底往上的，就像现实中的一个箱子。

3. 数据块的参数

数据块的参数如图 6-4 所示。

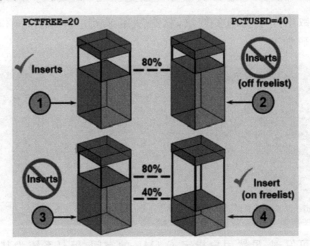

图 6-4 数据块的参数

下面就以图中的块参数为例，具体分析每个块是否允许插入数据的规则。

(1) 首先，观察块中各个参数的含义与值。

① pctfree：剩余空间的百分比小于或等于此参数，停止插入数据，图中对应的值为 20%。

② pctused：使用空间的百分比小于或等于此参数，可以插入数据，图中对应的值为 40%。

③ freelist：可以插入数据的状态。

(2) 插入规则分析。块头不在指定的百分比范围内，如果剩余空间大于 20%，这个块就纳入 freelist 中；如果要插入一条数据，freelist 就会扫描这个块，以检查其他是否可以存放要插入的数据。如果小于 20%，就说明这个块已经满了，会从 freelist 中摘除。注意，插入数据不作为扫描的对象。

如果一个小于20%空闲的块会从freelist上摘除,那么一个块在什么情况下会被重新挂到freelist上呢?对于一个已经从freelist上摘除的块,由于删除更新操作,其空间会得到释放,当占用空间小于40%时,也就是空闲空间大于60%时,这个块被认为是空闲的块,因此又会被重新挂到freelist上。

(3)数据块插入步骤。

① 只有数据块已使用空间比例小于80%才可有新的数据行插入,因为pctfree为20,意味着必须保留20%的剩余空间以用于更新已存在的数据。

② 块内的剩余空间将用于更新已存在的数据,不能插入新的数据行,直到块内已使用空间比例小于39%时。

③ 当块内已用空间低于40%时,方可插入新的行。

④ 回到步骤①。此过程往复循环。

4. 行链

如果块设置得过小,当要插入一数据行的长度大于块的时候,就需要串联多个块存储,称之为行链(Row Chaining),如图6-5所示。

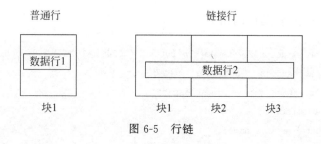

图6-5 行链

5. 行迁移

当一数据行因更新而需要增加数据长度时,如果存储此数据行的块已经没有足够的空间存放新增加的数据,就必须把整个数据行的内容迁移到另外一个块。称之为行迁移(Row Migration),如图6-6所示。

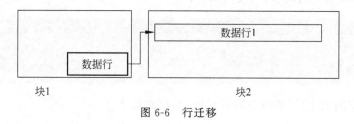

图6-6 行迁移

6. 数据块大小设置

怎样确定块大小呢?若要设置其大小,就需要考虑以下两个因素。

(1)数据库环境类型。在联机分析系统(OLAP)或决策支持系统(DSS)下,用户需要完成许多运行时间很长的查询操作,所以应当使用大的块。在联机事务处理过程(OLTP)中,如果用户想处理大量的小型事务,则建议采用较小的块,以便获得更好的效果。

(2) 数据库缓冲区的大小。数据库缓冲区的大小由块的大小和初始化文件的 db_block_buffers 参数决定。其大小最好设为操作系统 I/O 的整数倍。

6.6 项目案例

6.6.1 对表空间相关的操作实例

视频讲解

表空间操作步骤及其代码如下所述。

步骤 1　创建表空间 DLUFL,其代码如下：

```
SQL>create tablespace dlufl
  2   datafile 'D:\ORACLE19C\ORADATA\NEWORCL\DLUFL01.DBF'
  3   size 10m autoextend on;


表空间已创建.
```

步骤 2　查询表空间,其代码如下：

```
SQL>select t.tablespace_name, round(sum(bytes / (1024 * 1024)), 0) ts_size
  2   from dba_tablespaces t, dba_data_files d
  3   where t.tablespace_name = d.tablespace_name
  4   and   t.tablespace_name = 'dlufl'
  5   group by t.tablespace_name;


TABLESPACE     TS_SIZE
----------     -------
DLUFL              10
```

步骤 3　修改表空间大小,其代码如下：

```
SQL>alter database datafile 'D:/ORACLE19C/ORADATA/NEWORCL/DLUFL01.DBF' resize 80m;


数据库已更改.
```

在修改表空间大小后,继续查看表空间,其代码如下：

```
SQL>select t.tablespace_name, round(sum(bytes / (1024 * 1024)), 0) ts_size
  2   from dba_tablespaces t, dba_data_files d
  3   where t.tablespace_name = d.tablespace_name
  4   and   t.tablespace_name = 'dlufl'
  5   group by t.tablespace_name;
```

```
TABLESPACE     TS_SIZE
---------- ----------
DLUFL              80
```

步骤 4　添加文件至表空间,其代码如下:

```
SQL>alter tablespace dlufl add datafile  'D:/ORACLE19C/ORADATA/NEWORCL/DLUFL02.DBF' size 50m autoextend on;

表空间已更改.
```

步骤 5　更改表空间权限,其代码如下:

```
SQL>alter tablespace dlufl read only:

表空间已更改.
```

向只读表空间中插入数据文件,观察插入效果,其代码如下:

```
SQL>alter tablespace  dlufl add datafile  'D:/ORACLE19C/ORADATA/NEWORCL/DLUFL03.DBF' size 50m autoextend on;
alter tablespace  dlufl add datafile  'D:/ORACLE19C/ORADATA/NEWORCL/DLUFL03.DBF' size 50m autoextend on
                  *
第 1 行出现错误:
ORA-01641:表空间 'DLUFL' 未联机 - 无法添加数据文件
```

步骤 6　删除表空间,其代码如下:

```
SQL>drop tablespace dlufl including contents and datafiles;

表空间已删除.
```

步骤 7　删除操作后查看表空间,其代码如下:

```
SQL>select tablespace_name from dba_tablespaces;

TABLESPACE
----------
SYSTEM
SYSAUX
UNDOTBS1
TEMP
USERS
```

6.6.2 用户与表空间操作实例

新建一个用户 u1,并利用代码帮助其实现为特定用户分配指定表空间。

步骤 1　创建用户 u1,其代码如下:

```
SQL>create user u1 identified by u1;

用户已创建。
```

步骤 2　观察该用户的默认表空间分配情况,其代码如下:

```
SQL>select username,default_tablespace from dba_users order by username;

USERNAME                        DEFAULT_TABLESPACE
------------------------------  ------------------
ANONYMOUS                       SYSAUX
APPQOSSYS                       SYSAUX
AUDSYS                          USERS
CTXSYS                          SYSAUX
HR                              SYSAUX
SYS                             SYSTEM
T1                              USERS
……
已选择 38 行。
```

步骤 3　创建表空间,使用默认表空间,其代码如下:

```
SQL>create tablespace ts1 datafile 'D:\ORACLE19C\ORADATA\NEWORCL\TS1DATA.DBF' size 10m;

表空间已创建。
```

步骤 4　更改用户的默认表空间为 TS1,其代码如下:

```
SQL>alter user u1 default tablespace ts1;

用户已更改。
```

步骤 5　为用户 u1 授权,其代码如下:

```
SQL>grant create session,create table,create view,create sequence,unlimited tablespace to u1;

授权成功。
```

步骤 6　以用户 u1 登录,查询其权限,其代码如下:

```
SQL> conn u1/u1;
已连接.
SQL> select * from session_privs;

PRIVILEGE
----------------------------------------
CREATE SEQUENCE
CREATE VIEW
CREATE TABLE
UNLIMITED TABLESPACE
CREATE SESSION
```

步骤 7　以 sys 登录,级联删除用户 u1,其代码如下:

```
SQL> conn sys as sysdb
输入口令:
已连接.
SQL> drop user u1 cascade;

用户已删除.
```

步骤 8　级联删除新创建的表空间,其代码如下:

```
SQL> drop tablespace ts1  including contents and datafiles;

表空间已删除.
```

6.6.3　表段与块操作实例

步骤 1　查看数据库相关情况,包括查看数据库的版本以及查看数据库的创建日期和归档方式。

(1) 查看数据库的版本,其代码如下:

```
SQL> select version from product_component_version where substr(product, 1, 6) = 'Oracle';

VERSION
--------------------------------
19.0.0.0.0
```

(2) 查看数据库的创建日期和归档方式,其代码如下:

```
SQL> select created, log_mode, log_mode from v$database;

CREATED        LOG_MODE     LOG_MODE
----------     ----------   ----------
25-10月 -19    ARCHIVELOG   ARCHIVELOG
```

步骤 2 创建表空间 UNIFORMTB,其代码如下:

```
SQL> create tablespace uniformtb
  2    datafile  'D:/ORACLE19C/ORADATA/NEWORCL/UNIFORMTB01.DBF'
  3    size 25m autoextend on next 1280K maxsize 1000m,
  4    'D:/ORACLE19C/ORADATA/NEWORCL/UNIFORMTB02.DBF'
  5    size 25m autoextend on next 1280K maxsize 1000m
  6    extent management local uniform size 64K;

表空间已创建.
```

步骤 3 查看表空间情况,包括查看当前表空间和表空间的空间分配情况。
(1) 查看当前表空间,其代码如下:

```
SQL> select * from v$tablespace;

       TS# NAME                    INC BIG FLA ENC    CON_ID
---------- ----------------------- --- --- --- ---    ------
         1 SYSAUX                  YES NO  YES             0
         0 SYSTEM                  YES NO  YES             0
         2 UNDOTBS1                YES NO  YES             0
         4 USERS                   YES NO  YES             0
         3 TEMP                    NO  NO  YES             0
         7 UNIFORMTB               YES NO  YES             0

已选择 6 行.
```

或执行如下代码:

```
SQL> select tablespace_name,status,contents from dba_tablespaces;

TABLESPACE   STATUS    CONTENTS
----------   -------   -------------------
SYSTEM       ONLINE    PERMANENT
SYSAUX       ONLINE    PERMANENT
```

```
UNDOTBS1        ONLINE       UNDO
TEMP            ONLINE       TEMPORARY
USERS           ONLINE       PERMANENT
UNIFORMTB       ONLINE       PERMANENT
```

已选择 6 行.

(2) 查看表空间的空间分配情况。

① 查看已用表空间的名称及大小,其代码如下:

```
SQL> select t.tablespace_name, round(sum(bytes / (1024 * 1024)), 0) ts_size from dba_
tablespaces t, dba_data_files d  where t.tablespace_name = d.tablespace_name  group by t.
tablespace_name;

TABLESPACE       TS_SIZE
----------       -------
SYSTEM             910
UNIFORMTB           50
SYSAUX             720
UNDOTBS1            65
USERS                5
```

② 查看空闲表空间的大小,其代码如下:

```
SQL> select sum(bytes)/(1024 * 1024) as free_space,tablespace_name from dba_free_space group
by tablespace_name;

FREE_SPACE    TABLESPACE
----------    ----------
     7.25     SYSTEM
       48     UNIFORMTB
   58.625     SYSAUX
   36.375     UNDOTBS1
   2.3125     USERS
```

(3) 查看表空间中物理文件,其代码如下:

```
SQL> col name format a50
SQL> select file#, ts#,name  from v$datafile;

    FILE#         TS#    NAME
----------    -------   -----------------------------------
```

```
         1        0   D:\ORACLE19C\ORADATA\NEWORCL\SYSTEM01.DBF
         3        1   D:\ORACLE19C\ORADATA\NEWORCL\SYSAUX01.DBF
         4        2   D:\ORACLE19C\ORADATA\NEWORCL\UNDOTBS01.DBF
         5        7   D:\ORACLE19C\ORADATA\NEWORCL\UNIFORMTB01.DBF
         7        4   D:\ORACLE19C\ORADATA\NEWORCL\USERS01.DBF
         8        7   D:\ORACLE19C\ORADATA\NEWORCL\UNIFORMTB02.DBF
```

已选择 6 行。

```
SQL>col  to_char(file_id) format a10
SQL>select tablespace_name,to_char( file_id), file_name, round(bytes / (1024 * 1024), 0)
total_space from dba_data_files order by tablespace_name;

TABLESPACE      TO_CHAR(FI   FILE_NAME                                           TOTAL_SPACE
-----------     ----------   ------------------------------------------------    -----------
SYSAUX          3            D:\ORACLE19C\ORADATA\NEWORCL\SYSAUX01.DBF                   720
SYSTEM          1            D:\ORACLE19C\ORADATA\NEWORCL\SYSTEM01.DBF                   910
UNDOTBS         14           D:\ORACLE19C\ORADATA\NEWORCL\UNDOTBS01.DBF                   65
UNIFORMTB       8            D:\ORACLE19C\ORADATA\NEWORCL\UNIFORMTB02.DBF                 25
UNIFORMTB       5            D:\ORACLE19C\ORADATA\NEWORCL\UNIFORMTB01.DBF                 25
USERS           7            D:\ORACLE19C\ORADATA\NEWORCL\USERS01.DBF                      5
```

已选择 6 行。

步骤 4 在该表空间上创建表 test，其代码如下：

```
SQL>create table test(
  2   id number,
  3   name varchar2(20)
  4   )
  5   tablespace uniformtb
  6   storage( initial 1m next 512k minextents 5 pctincrease 0);
```

表已创建。

步骤 5 查询与该表相关的存储参数。
(1) 查看在创建表后，该表空间的空闲空间大小，其代码如下：

```
SQL>select sum(bytes)/(1024 * 1024) as free_space,tablespace_name from dba_free_space group
by tablespace_name;

FREE_SPACE  TABLESPACE
----------  ----------
      7.25  SYSTEM
```

```
    45   UNIFORMTB
50.625   SYSAUX
36.375   UNDOTBS1
2.3125   USERS
```

从步骤 3 可以获知表空间 uniformtb 的空闲空间为 48MB,而上述结果中表空间 uniformtb 的空闲空间为 45MB。即 48－45＝3,说明表 test 占用了 3MB 空间,这是为什么呢？带着疑问继续执行下列操作。

(2) 查询表 test 段信息,其代码如下：

```
SQL>select table_name, initial_extent, next_extent, min_extents, max_extents, pct_increase
    from user_tables where table_name = 'TEST';

TABLE_NAME   INITIAL_EXTENT   NEXT_EXTENT   MIN_EXTENTS   MAX_EXTENTS   PCT_INCREASE
----------   --------------   -----------   -----------   -----------   ------------
TEST         3145728          524288        1             2147483645    0
```

从此操作结果中发现,initial_extent 为 3 145 728（3MB）,next_extent 为 524 288 (512KB)。

(3) 再次查询表 test 信息,其代码如下：

```
SQL>select segment_name, segment_type, blocks, extents, tablespace_name from DBA_segments
    where segment_name = 'TEST';

SEGMENT_NAME   SEGMENT_TYPE   BLOCKS   EXTENTS   TABLESPACE
------------   ------------   ------   -------   ----------
TEST           TABLE          384      48        UNIFORMTB
```

从此操作结果中发现,blocks 为 384,extents 为 48。这些代表什么含义呢？带着疑问继续执行下列操作。

步骤 6　查询表 test 分配的区数。

```
SQL>select count(*) from dba_extents where segment_name = 'TEST';
  COUNT(*)
----------
        48
```

可以通过 dba_extents 视图观察表 test 中区的分布情况。

```
SQL>select file_id, extent_id, block_id, blocks from dba_extents
  2  where owner = 'SYS' and segment_name = 'TEST' order by 2;
```

```
    FILE_ID   EXTENT_ID  BLOCK_ID   BLOCKS
   -------- ---------- ---------- ------
        5          0        128        8
        8          1        128        8
        5          2        136        8
        8          3        136        8
        5          4        144        8
        8          5        144        8
        5          6        152        8
        8          7        152        8
        5          8        160        8
        8          9        160        8
        5         10        168        8
.......

已选择 48 行.
```

从结果中发现,查询结果 48 代表表 test 段占用 48 个区(Extent),每个区又包含 8 个块。这就解释了上个结果中的 48 和 384 的含义。

步骤 7　查询与块相关的信息,其代码如下:

```
SQL> select name, value from v$parameter where name = 'db_block_size';

NAME                                               VALUE
-------------------------------------------------- ----------
db_block_size                                      8192
```

视频讲解

6.6.4　数据库行链操作实例

步骤 1　创建表空间 STUDENT,其代码如下:

```
SQL> create tablespace student
  2    datafile 'D:\ORACLE19C\ORADATA\NEWORCL\STUDENT01.DBF'
  3    size 25m autoextend on next 1280k maxsize 1000m
  4    extent management local;

表空间已创建。
```

步骤 2　创建必要的检测行链的表,其代码如下:

```
SQL> @D:\SOFTWARE\ORACLE19C\RDBMS\ADMIN\UTLCHAIN.SQL

表已创建。
```

分析表并检测行链现象,其代码如下:

```
SQL> select table_name, head_rowid from chained_rows;
```

未选定行

步骤 3 使行迁移及出现行迁移现象的表,其代码如下:

```
SQL> create table test1 ( x int primary key,
  2    a char(2000), b char(2000), c char(2000),
  3    d char(2000), e char(2000) ) tablespace student;
```

表已创建.

步骤 4 插入数据,使其占用 10000B 以上的存储空间,其代码如下:

```
SQL> insert into test1 values(1, 'test','test','test','test','test');
```

已创建 1 行.

步骤 5 分析该表,发现行链现象(因为 10000＞8192,所以出现行链现象)。

```
SQL> analyze table test1 list chained rows;
```

表已分析.
```
SQL> col table_name format a20
SQL> select table_name,head_rowid from chained_rows;

TABLE_NAME           HEAD_ROWID
-------------------- --------------------
TEST1                AAASgWAACAAAACGAAA
```

6.6.5 数据库行迁移操作实例

视频讲解

步骤 1 创建表 S,其代码如下:

```
SQL> create table s(sno char(10) primary key,
  2    sname char(990),
  3    ssex char(500))
  4    pctfree 5
  5    pctused 40
  6    tablespace student;
```

表已创建.

步骤2 持续插入8行数据,且只给这8行数据的前两列赋值,第3列为空,这样每个数据行占有1000B。预测在插入前7行数据后,第8行数据会插入到下一个块中。这是因为块大小为8192B,需要预留pctfree值规定的空闲空间,约为410B(8192×5%)。这样剩余的可以容纳数据的大小为8192－410＝7782B。而前7行数据将占有7000B,第8个数据将插入到下一个块中。代码如下:

```
SQL> insert into s(sno,sname) values('0201','name1');
已创建 1 行.
SQL> insert into s(sno,sname) values('0202','name2');
已创建 1 行.
SQL> insert into s(sno,sname) values('0203','name3');
已创建 1 行.
SQL> insert into s(sno,sname) values('0204','name4');
已创建 1 行.
SQL> insert into s(sno,sname) values('0205','name5');
已创建 1 行.
SQL> insert into s(sno,sname) values('0206','name6');
已创建 1 行.
SQL> insert into s(sno,sname) values('0207','name7');
已创建 1 行.
SQL> insert into s(sno,sname) values('0208','name8');
已创建 1 行.
```

步骤3 查询刚创建的8个数据行,显示sno字段和rowid的相关信息,其代码如下:

```
SQL> select sno,dbms_rowid.rowid_relative_fno(rowid)
file#,dbms_rowid.rowid_block_number(rowid) block#,dbms_rowid.rowid_row_number(rowid) row
# from s;

SNO        FILE#       BLOCK#      ROW#
---------- ----------- ----------- -------
0201       2           149         0
0202       2           149         1
0203       2           149         2
0204       2           149         3
0205       2           149         4
0206       2           149         5
0207       2           149         6
0208       2           150         0

已选择 8 行.
```

步骤4 用update语句对插入的数据行中的ssex列进行更新。第149号数据块在插入 7 行数据后,它剩余的空间为 8192－7000＝1192B。这么多字节能容纳两个数据行 ssex 字段的 update 操作。因此,推测在第 3 个数据行后将陆续出现行迁移。

```
SQL> update s set ssex = 'F';

已更新 8 行.
```

步骤 5　出现行迁移现象,并从 chained_rows 表中查看出现迁移的数据行,以印证步骤 4 的推测。

```
SQL> analyze table s list chained rows;

表已分析.
SQL> select table_name,
  2   dbms_rowid.rowid_relative_fno(head_rowid) file#,
  3   dbms_rowid.rowid_block_number(head_rowid) block#,
  4   dbms_rowid.rowid_row_number(head_rowid) row# from chained_rows
  5   ;

TABLE_NAME        FILE#     BLOCK#      ROW#
-----------   ---------   ---------  ---------
TEST1               2         134        0
S                   2         149        2
S                   2         149        4
```

步骤 6　再次进行步骤 3 的查询,其代码如下:

```
SQL> select sno,dbms_rowid.rowid_relative_fno(rowid)
file#,dbms_rowid.rowid_block_number(rowid) block#,dbms_rowid.rowid_row_number(rowid) row
# from s;

SNO         FILE#     BLOCK#      ROW#
---------  --------  --------   ---------
0201          2         149         0
0202          2         149         1
0203          2         149         2
0204          2         149         3
0205          2         149         4
0206          2         149         5
0207          2         149         6
0208          2         150         0

已选择 8 行.
```

从查询结果中发现，其结果相同。这是因为两行数据迁移的数据行的 row♯ 的值没有变。所以在访问这行数据时，至少需要读两个块，这就降低了效率。所以如果出现行迁移现象比较严重的情况，就需要将相关行删除掉并重新插入。步骤 7～步骤 10 为其操作方法。

步骤 7　创建临时表存储迁移的数据行，其代码如下：

```
SQL> create table s_temp as
  2  select * from s
  3  where rowid in
  4  (select head_rowid from chained_rows where table_name = 'S');

表已创建.
```

步骤 8　将迁移的数据行从表 s 中删除，其代码如下：

```
SQL> delete from student
  2  where rowid IN
  3  (select head_rowid
  4  from chained_rows
  5  where table_name = 'S');

已删除 2 行.
```

步骤 9　从临时表中重新插入数据，并将临时表删除掉，其代码如下：

```
SQL> insert into s select * from s_temp;

已创建 2 行.
SQL> drop table s_temp;

表已删除.
```

步骤 10　重新查询表 s 的 rowid 情况，发现前面迁移的两行数据的 block♯ 发生了改变，其代码如下：

```
SQL> select sno,dbms_rowid.rowid_relative_fno(rowid)
file♯,dbms_rowid.rowid_block_number(rowid) block♯,dbms_rowid.rowid_row_number(rowid) row
♯ from s;

SNO          FILE♯       BLOCK♯      ROW♯
---------- ---------- ---------- ----------
0201            2          149         0
0202            2          149         1
0203            2          150         4
```

```
0204            2       149     3
0205            2       150     3
0206            2       149     5
0207            2       149     6
0208            2       150     0
```

已选择 8 行.

第7章

Oracle 网络配置管理

PPT 视频讲解

7.1 Oracle 数据库网络配置

Oracle 数据库的网络配置既简单又复杂。学习时由简入繁,只要简单的弄透彻了,繁杂的也就无非是时间问题了。

7.1.1 Oracle 网络服务组件

无论是应用程序,还是使用 SQL Plus 工具连接远程数据库,都需要建立客户端与数据库服务器之间的一个 Oracle 连接。Oracle 提供了 Oracle Net Services 组件,用于方便地配置和管理网络连接。

Oracle Net Services 组件由 Oracle Net、Oracle Net 监听器和网络配置工具 3 个部分组成。

1. Oracle Net

Oracle Net 是同时驻留在 Oracle 服务器和客户端上的一个软件层,它负责建立与维护客户端应用程序到数据库服务器的连接。

2. Oracle Net 监听器

Oracle Net 监听器是位于服务器端的一个后台进程,它负责对客户端传出的连接请求进行监听,并且负责对服务器端的连接负荷进行调整。当客户端试图建立一个到服务器端的网络会话时,首先是由监听器处理实际的网络连接请求。一旦客户端与服务器的连接已经建立,客户端和服务器即可直接通信,不需要监听器的参与。

3. 网络配置工具

Oracle 同时提供图形化界面和命令行方式的网络配置工具,主要的 3 个网络配置工具如下。

(1) Oracle Net Configuration Assistant 图形化工具：通常在完成 Oracle 数据库服务器时会自动启动，利用它可完成基本的网络（监听器）的配置工作。

(2) 命令行配置工具 LSNRCTL：对监听器进行配置、管理与监视。

(3) Oracle Net Manager 图形化管理工具：提供对 Oracle 所有网络组件进行详细配置的集中化管理界面。DBA 可对创建的监听器进行进一步的细致调整和配置。

7.1.2 Oracle 网络连接的流程

在客户端的用户进程可以发出连接指令，具体的连接指令有很多，后续会相继对其进行阐述。在这里首先了解 Oracle 网络连接的流程，如图 7-1 所示。

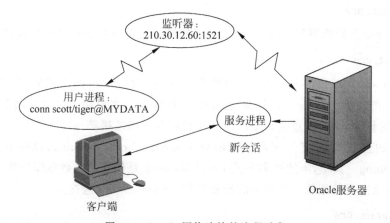

图 7-1　Oracle 网络连接的流程示意

Oracle 网络连接的流程描述如下。

(1) 客户端发起连接。在这个过程中需要服务器、端口、协议、数据库服务名。

(2) 客户端一旦与监听器建立连接，则在客户端生成用户进程，同时监听器会判断客户端所请求的服务名是否是自己所管理的服务名。如果客户端传过来的连接字符串不包含服务名，则报错；如果请求的服务名不是自己管理的，则报错并中断；如果请求的服务名是自己管理的，监听器就在数据库服务器上创建服务器进程。

(3) 监听器在创建服务器进程以后，会将用户进程与服务器进程建立连接，之后监听器退出与客户端的连接。

(4) 服务器进程根据用户进程提供的用户名和密码到数据字典里判断是否正确。

(5) 如果用户名和密码不匹配，则报错；如果匹配，则分配 PGA，并生成会话。

7.1.3 Oracle 网络配置文件

Oracle 产品安装完成后，服务器端和客户端都需要进行网络配置才能实现网络连接，即服务器端配置监听器，客户端配置网络服务名。具体网络配置参数文件如图 7-2 所示。

Oracle Net Services 的配置文件有 listener.ora、sqlnet.ora 和 tnsnames.ora 3 个。

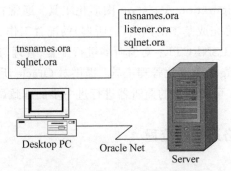

图 7-2　网络配置参数文件

1. listener.ora

listener.ora 为监听器配置文件,位于服务器端。它监听从客户发来的连接请求。

2. sqlnet.ora

sqlnet.ora 为网络概要配置文件,主要存储一些通信参数,如认证方式、安全设置等。sqlnet.ora 既可以位于 Oracle 服务端,也可以位于 Oracle 客户端。它主要用于控制客户端和服务器端的行为,例如设置会话跟踪级别和控制客户端连接等。

数据库的连接方式有很多种,如 Easy Connect、Local Naming、Directory Naming、External Naming。其中,本地命名方式 Local Naming 是较常见的数据库连接方式,如果选择了该方式,就需要使用位于客户端的配置文件 tnsnames.ora 连接。

3. tnsnames.ora

tnsnames.ora 为本地 Net 服务名配置文件,主要存放连接 Oracle 服务器的一些必须信息,如通信协议、服务器主机、端口号和服务名。

综上所述,Oracle 网络配置有 listener.ora、sqlnet.ora 和 tnsnames.ora 3 个配置文件,其目录均在 $Oracle_HOME/NETWORK/ADMIN 中。Oracle Net Configuration Assistant 和 Oracle Net Manager 工具都能用来配置监听器和网络服务名。

7.2　网络概要配置

网络概要配置是指客户端与服务器端通信时的命名方式、认证方式、安全设置等配置。设置后的配置信息将体现在其对应的参数文件为 sqlnet.ora 中,该文件位于客户端和服务器端,主要用于主机命名法和目录命名法。通过这个文件决定如何找到连接中出现的连接字符串。

7.2.1　网络概要配置步骤

通过 Oracle 提供的配置工具可以进行网络概要配置,具体步骤如下。

步骤 1　用 netca 命令或在 Oracle 19c 的配置和移植工具中选择"Oracle Net Configuration Assistant:欢迎使用"对话框,选择"命名方法配置"选项,单击"下一步"按钮,如图 7-3 所示。

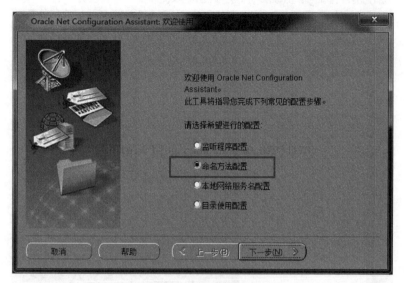

图 7-3　命名方法配置

步骤 2　在弹出的"Oracle Net Configuration Assistant 命名方法配置,请选择命名方法"对话框中选择"本地命名"和"轻松连接命名"选项,如图 7-4 所示。

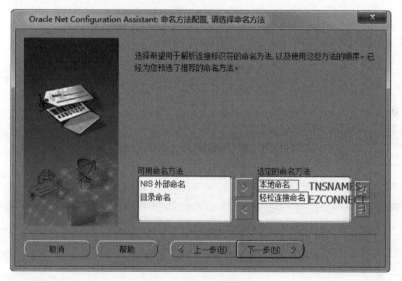

图 7-4　选择命名方法

由于在可用命名法中选择了"本地命名"选项,所以后续需要使用 tnsnames.ora 配置文件进行数据库连接。

步骤 3　配置结束后,可以在"Oracle Net Manager"窗口中查看概要文件配置相关信息,也可以直接在 Oracle Net Manager 中进行命名方法配置,如图 7-5 所示。

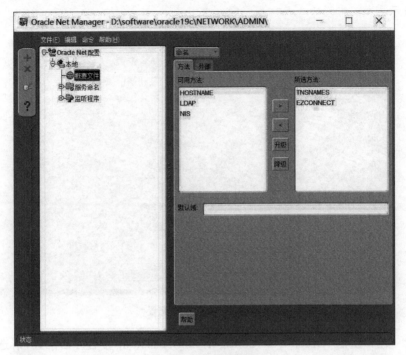

图 7-5　命名方法配置

7.2.2　网络概要配置文件——sqlnet.ora

按照 7.2.1 节的网络概要配置操作步骤，打开 sqlnet.ora 文件，该文件的配置如图 7-6 所示。

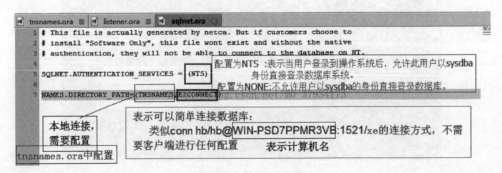

图 7-6　sqlnet.ora 文件

假如此时客户端输入"sqlplus sys/Oracle @neworcl"，那么系统会首先查找 sqlnet.ora 的网络配置信息；如果文件配置信息为 sqlnet.authentication_services=(nts) names.directory_path=(tnsnames,hostname)，那么，客户端就会首先在 tnsnames.ora 文件中找 neworcl 的记录；如果没有相应的记录，就尝试把 neworcl 当作一个主机名，并通过网络途径解析它的 IP 地址。然后连接这个 IP 上的 global_dbname=neworcl 实例。当然，这里的 neworcl 并不是一个主机名。

如果 names.directory_path=（tnsnames），那么客户端就只会从 tnsnames.ora 查找 neworcl 的记录，括号中还有其他选项，如 LDAP 等并不常用。

7.3 服务器端网络配置——监听器

监听是客户端连接到数据库的必经之路。监听器（Listener）是 Oracle 服务器端的一种网络服务，主要用于监听客户端向数据库服务器端发出的连接请求。它通常只存在于数据库服务器端，它的设置也是在服务器端完成的。Listener 的示意如图 7-7 所示。

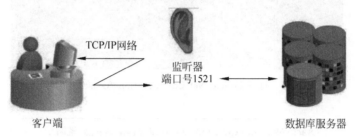

图 7-7 Listener 的示意

如果计算机有 Oracle 服务器端监听服务，那么打开 Windows 的服务页面，就可以找到 Oracle 的监听服务，如图 7-8 所示。

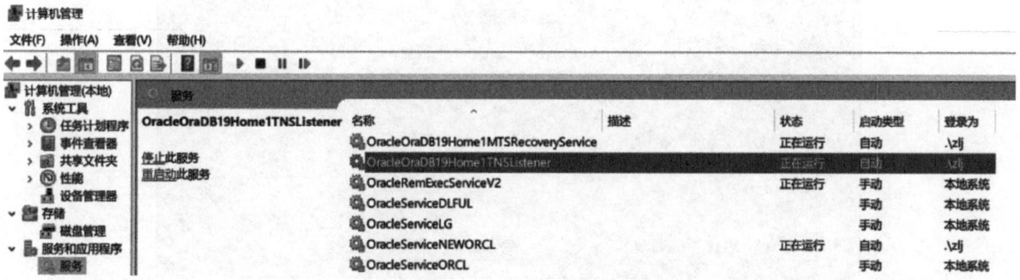

图 7-8 监听服务

7.3.1 监听器的配置

服务器端监听器配置信息包括监听协议、地址等相关信息。配置信息保存在名为 listener.ora 的文件中。可以手动编辑 listener.ora 监听器文件，当然还可以利用 Oracle Net Configuration Assistant 和 Oracle Net Manager 工具配置监听器和网络服务名。

下面通过 Oracle Net Configuration Assistant 演示监听器的配置过程。

步骤 1　打开 Oracle Net Configuration Assistant 向导，在"Oracle Net Configuration Assistant：欢迎使用"对话框选择"监听程序配置"选项，单击"下一步"按钮，如图 7-9 所示。

步骤 2　在弹出的对话框中选择"添加"选项，单击"下一步"按钮，如图 7-10 所示。

步骤 3　在弹出的对话框中，在"监听程序名"文本框中输入程序名，然后单击"下一步"按钮，如图 7-11 所示。

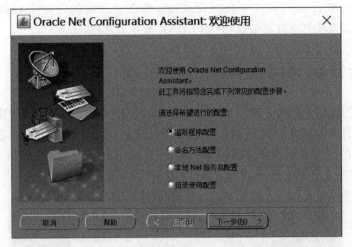

图 7-9　打开 Oracle Net Configuration Assistant 向导

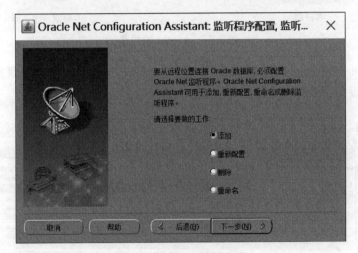

图 7-10　选择"添加"选项

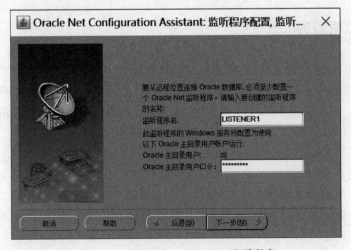

图 7-11　配置一个 Oracle Net 监听程序

步骤 4 在弹出的对话框中选择协议配置监听程序,接受连接,然后单击"下一步"按钮,如图 7-12 所示。

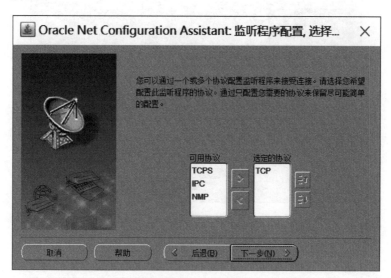

图 7-12 选择协议

步骤 5 在弹出的对话框中选择"请使用另一个端口号"选项,并在其文本框中输入端口号,然后单击"下一步"按钮,如图 7-13 所示。

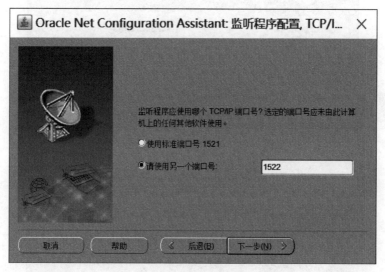

图 7-13 选择端口

步骤 6 在弹出的对话框中选择"否"选项,单击"下一步"按钮,如图 7-14 所示。
步骤 7 在弹出的对话框中选择默认的监听程序,单击"下一步"按钮,如图 7-15 所示。
步骤 8 在弹出的"监听程序配置完成"对话框中单击"下一步"按钮即可,如图 7-16 所示。

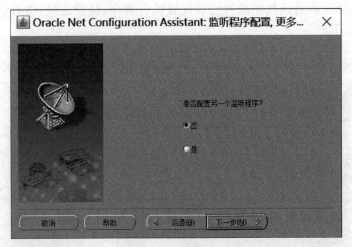

图 7-14 设置是否配置另一个监听程序

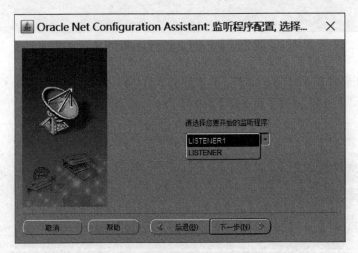

图 7-15 选择要开始的监听程序

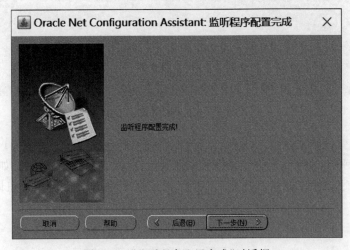

图 7-16 "监听程序配置完成"对话框

步骤 9　配置结束后,选择在"计算机管理"→"服务"选项,即可发现该监听服务,如图 7-17 所示。

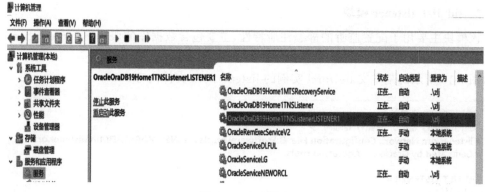

图 7-17　监听服务

在监听器配置完毕之后,会在服务管理处新增加一个 Listener1 的服务,该服务就是一个新的监听器,监听的端口是 1522(默认是 1521),如果用户想利用新的监听器进行数据库服务器的连接,这时候需要重新配置新的网络服务名,使用新的端口 1522 才能进行数据库连接。

7.3.2　监听配置文件——listener.ora

以下是 Oracle 19c 在 sample 文件夹中提供的一个 listener.ora 配置文件的模板。

```
LISTENER =
  (ADDRESS_LIST =
    (ADDRESS = (PROTOCOL = tcp)(HOST = localhost)(PORT = 1521))
    (ADDRESS = (PROTOCOL = ipc)(KEY = PNPKEY)))
SID_LIST_LISTENER =
  (SID_LIST =
  (SID_DESC =
      #BEQUEATH CONFIG
      (GLOBAL_DBNAME = salesdb.mycompany)
      (SID_NAME = sid1)
      (ORACLE_HOME = /private/app/oracle/product/8.0.3)
      #PRESPAWN CONFIG
      (PRESPAWN_MAX = 20)
    (PRESPAWN_LIST =
       (PRESPAWN_DESC = (PROTOCOL = tcp)(POOL_SIZE = 2)(TIMEOUT = 1))
    )
  )
)
```

从以上的配置模板可以看到,该监听配置文件有固定的格式,主要分为 listener 和 sid_list_listener 两大模块。

1. listener 模块

该模块主要包含监听名字、连接协议、监听主机和监听端口等监听的基本配置信息。

2. sid_list_listener 模块

该模块主要用于配置监听的静态注册特性，主要包含数据库的服务名 Oracle_home、实例名等信息。

图 7-18 所示的是本文 neworcl 实例的 listener. ora 文件内容。

```
listener.ora - 记事本
文件(F) 编辑(E) 格式(O) 查看(V) 帮助(H)
# listener.ora Network Configuration File: D:\software\oracle19c\NETWORK\ADMIN\listener.ora
# Generated by Oracle configuration tools.

SID_LIST_LISTENER =
 (SID_LIST =
  (SID_DESC =
   (SID_NAME = CLRExtProc)
   (ORACLE_HOME = D:\software\oracle19c)
   (PROGRAM = extproc)
   (ENVS = "EXTPROC_DLLS=ONLY:D:\software\oracle19c\bin\oraclr19.dll")
  )
 )

LISTENER =
 (DESCRIPTION_LIST =
  (DESCRIPTION =
   (ADDRESS = (PROTOCOL = TCP)(HOST = localhost)(PORT = 1521))
   (ADDRESS = (PROTOCOL = IPC)(KEY = EXTPROC1521))
  )
 )
第 22 行，第 1 列    100%    Windows (CRLF)    UTF-8
```

图 7-18 listener. ora

7.3.3 监听器的管理

1. 监听器操作命令

监听的常用命令包含启动关闭监听、查看监听状态、查看监听服务和重新装载监听服务等。用户可以在命令提示符中输入 LSNRCTL 后查看帮助。帮助中会罗列出监视器的管理操作命令，如 start、stop、status 等。

```
C:\WINDOWS\system32>LSNRCTL

LSNRCTL for 64 - bit Windows: Version 19.0.0.0.0 - Production on 16 - 2 月 - 2020 13:43:56

Copyright (c) 1991, 2019, Oracle.  All rights reserved.

欢迎来到 LSNRCTL，请键入"help"以获得信息.
```

```
LSNRCTL> help
以下操作可用

星号（*）表示修改符或扩展命令：

start        stop         status       services
servacls     version      reload       save_config
trace        quit         exit         set *
show *
```

其中常用的命令有以下 4 种。
(1) 获取所有监听器操作命令：

```
LSNRCTL> help
```

(2) 获取 start 命令的使用信息：

```
LSNRCTL> help start
```

(3) 获取监听进程当前的状态：

```
LSNRCTL> status
```

(4) 启动 Oracle 系统的默认监听进程：

```
LSNRCTL> start
```

2. 监听器的注册

在以上 4 种命令中，使用 lsnrctl status 命令可以查看某个服务是静态注册还是动态注册。注意，这里的实例名称 neworcl 和 CLRExtProc 与 listener.ora 中是一致的。仔细观察状态，如果是状态值为 UNKNOWN，表示实例是静态注册到监听；如果状态值是 READY，表示动态注册。动态注册和静态注册可以根据需要进行转换。

3. 监听日志

在上面的监听命令中很容易找到监听日志的路径，即监听程序日志文件：D:\ORACLE19C\DIAG\TNSLSNR\LAPTOP-8BM7LBRI\LISTENER\ALERT\LOG.XML。

通常在处理监听日志上有以下两种常用处理办法。
(1) 开启监听日志记录：

```
C:\WINDOWS\system32>lsnrctl set log_status on
```

(2) 关闭监听日志记录：

```
C:\WINDOWS\system32>lsnrctl set log_status off
```

7.4 客户端网络配置——网络服务名

7.4.1 网络服务名配置

安装了 Oracle 客户端之后就可以在以下目录找到 tnsname.ora 文件：$Oracle_HOME/NETWORK/ADMIN，如果未找到，就可以到其子目录 sample 的文件夹里查找。

对本地网络服务名的配置可以在客户端机器上使用 Oracle Net Configuration Assistant 或 Oracle Net Manager 图形配置工具对客户端进行配置，该配置工具实际上就是对 tnsnames.ora 文件的修改，所以也可以直接手动编辑修改 tnsnames.ora 文件。

下面介绍如何利用 Net Configuration Assistant 图形配置工具对客户端进行配置。其操作步骤如下。

步骤1 在"Oracle Net Configuration Assistant：欢迎使用"对话框中选择"本地 Net 服务名配置"选项，单击"下一步"按钮，如图 7-19 所示。

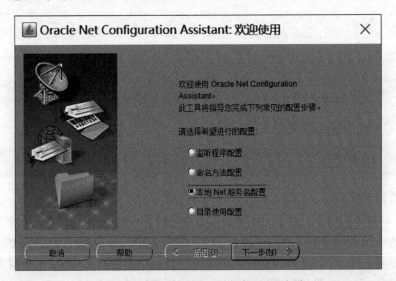

图 7-19 选择"本地 Net 服务名配置"选项

步骤2 在弹出的"Oracle Net Configuration Assistant：Net 服务名配置"对话框中选择"添加"选项，单击"下一步"按钮，如图 7-20 所示。

步骤3 在弹出的对话框中的"服务名"文本框中输入服务名，单击"下一步"按钮，如图 7-21 所示。

步骤4 在弹出的对话框中选择"TCP"选项，单击"下一步"按钮，如图 7-22 所示。

步骤5 在弹出的对话框中配置数据所在的主机名或 IP 地址和端口号，然后单击"下一步"按钮，如图 7-23 所示。

步骤6 在弹出的对话框中选择"是，进行测试"选项，然后单击"下一步"按钮，如图 7-24 所示。

第7章 Oracle网络配置管理

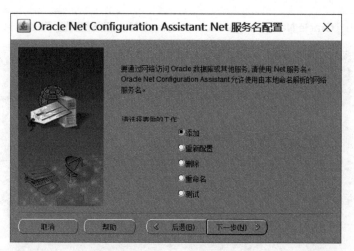

图 7-20 "Oracle Net Configuration Assistant：Net 服务名配置"对话框

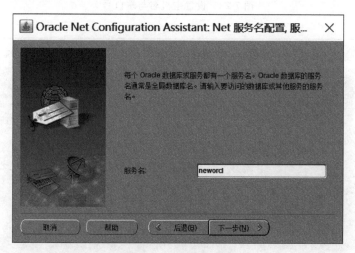

图 7-21 配置服务名

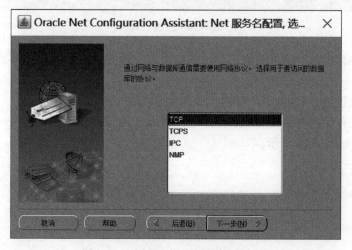

图 7-22 选择要访问的数据库的协议

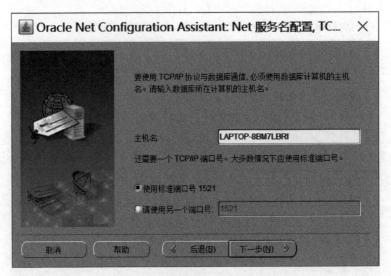

图 7-23 设置主机名与端口号

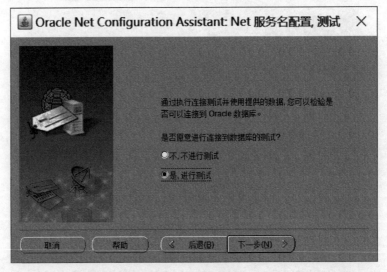

图 7-24 连接测试

步骤 7 在弹出的对话框中如果显示未连上,单击"更改登录"按钮,即可重新设置数据库用户名和密码,如图 7-25 所示。

步骤 8 配置服务名,单击"下一步"按钮,如图 7-26 所示。

步骤 9 在弹出的对话框中选择"否"选项,表示不需要配置另一个 Net 服务名,单击"下一步"按钮,如图 7-27 所示。

步骤 10 配置成功。

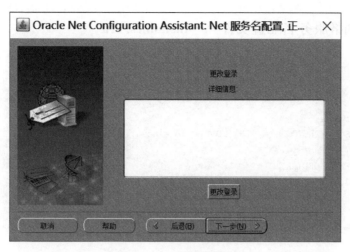

图 7-25 重置登录信息

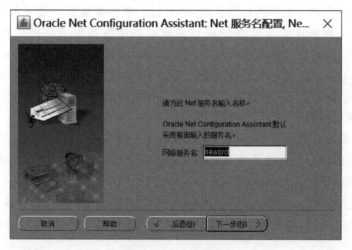

图 7-26 输入服务名

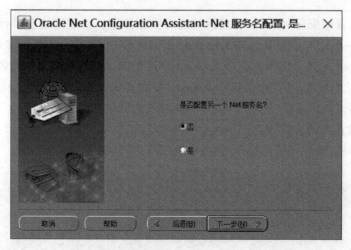

图 7-27 设置是否配置另一个 Net 服务名

7.4.2 网络服务名配置文件——tnsnames.ora

打开 tnsnames.ora 文件，其代码如下：

```
NEWORCL =
  (DESCRIPTION =
    (ADDRESS = (PROTOCOL = TCP)(HOST = localhost)(PORT = 1521))
    (CONNECT_DATA =
      (SERVER = DEDICATED)
      (SERVICE_NAME = neworcl)
    )
  )
```

其中：

PROTOCOL：客户端与服务器端通信的协议一般为 TCP。该内容一般不用改。

HOST：数据库监听所在的机器名或 IP 地址。数据库监听一般与数据库在同一个机器上，所以如果说数据库监听所在的机器一般也是指数据库所在的机器。在 UNIX 或 Windows 下，可以通过在数据库监听所在的机器的命令提示符下使用 hostname 命令得到机器名，或通过 ipconfig 命令得到 IP 地址。注意，不管是用机器名还是用 IP 地址，在客户端一定要用 ping 命令成功连通数据库所监听的机器；否则需要在 hosts 文件中加入数据库监听所在的机器名的解析。

PORT：数据库正在监听的端口。可以查看服务器端的 listener.ora 文件或在数据库监听所在的机器的命令提示符下通过 lnsrctl status[listener name] 命令查看。此处 PORT 的值一定要与数据库正在监听的端口一样。

SERVICE_NAME：在服务器端，用 system 用户名登录后，show parameter service_name 命令查看。

7.5 网络连接

网络基本配置结束后，便可以通过网络连接数据库了。以下列连接字符串为例：sqlplus sys/neworcl@neworcl as sysdba，系统首先查询 sqlnet.ora 文件，若名称解析方式为 tnsname，则查询 tnsnames.ora 文件。从该文件中查找 neworcl 的相关记录，进行主机名、端口、service_name 的比对。比对成功后，就可以建立与 Listener 进程的连接，并根据不同的服务器模式连接数据库服务器。至此，网络连接已经建立，Listener 进程的使命也就完成了。

具体连接数据库的方式可以归纳为以下 4 种。

（1）连接本机，操作系统认证，不需要 Listener 进程，其代码如下：

```
C:\WINDOWS\system32>set oracle_sid = neworcl
C:\WINDOWS\system32>sqlPlus / as sysdb

SQL Plus: Release 19.0.0.0.0 - Production on 星期日 2 月 16 14:35:15 2020
```

Version 19.3.0.0.0

Copyright (c) 1982, 2019, Oracle.   All rights reserved.

连接到:
Oracle Database 19c Enterprise Edition Release 19.0.0.0.0 - Production
Version 19.3.0.0.0

(2) 连接本机数据库。该方式不需要启动监听,而是需要启动数据库服务和输入用户名与密码,其代码如下:

```
C:\WINDOWS\system32>set oracle_sid = neworcl

C:\WINDOWS\system32>sqlPlus system/neworcl

SQL Plus: Release 19.0.0.0.0 - Production on 星期日 2 月 16 14:36:18 2020
Version 19.3.0.0.0

Copyright (c) 1982, 2019, Oracle.    All rights reserved.

上次成功登录时间:星期二 12 月 03 2019 12:47:11 +08:00

连接到:
Oracle Database 19c Enterprise Edition Release 19.0.0.0.0 - Production
Version 19.3.0.0.0
```

(3) 通过网络连接本机数据库,需要启动监听服务、数据库服务以及输入用户名和密码,其代码如下:

```
C:\WINDOWS\system32>set oracle_sid = neworcl

C:\WINDOWS\system32>sqlPlus system/neworcl

SQL Plus: Release 19.0.0.0.0 - Production on 星期日 2 月 16 14:36:18 2020
Version 19.3.0.0.0

Copyright (c) 1982, 2019, Oracle.    All rights reserved.

上次成功登录时间:星期二 12 月 03 2019 12:47:11 +08:00

连接到:
Oracle Database 19c Enterprise Edition Release 19.0.0.0.0 - Production
Version 19.3.0.0.0
```

(4) 通过远程连接数据库服务器,其代码如下:

```
C:\WINDOWS\system32>sqlplus hr/hr@neworcl

SQL Plus: Release 19.0.0.0.0 - Production on 星期日 2月 16 14:38:29 2020
Version 19.3.0.0.0

Copyright (c) 1982, 2019, Oracle.  All rights reserved.


ERROR:
ORA-12541: TNS: 无监听程序
```

如果此时未启动监听服务,则会出现错误提示。此时需要先启动监听,再进行远程连接数据库服务器,其代码如下:

```
C:\WINDOWS\system32>LSNRCTL

LSNRCTL for 64-bit Windows: Version 19.0.0.0.0 - Production on 16-2月 -2020 14:39:04
Copyright (c) 1991, 2019, Oracle.  All rights reserved.
欢迎来到LSNRCTL,请键入"help"以获得信息.
.......
LSNRCTL> start
启动 tnslsnr: 请稍候...
命令执行成功
LSNRCTL> exit
C:\WINDOWS\system32>set oracle_sid=neworcl
C:\WINDOWS\system32>sqlPlus hr/hr@neworcl

SQL Plus: Release 19.0.0.0.0 - Production on 星期日 2月 16 14:40:05 2020
Version 19.3.0.0.0

Copyright (c) 1982, 2019, Oracle.  All rights reserved.
上次成功登录时间: 星期日 2月  16 2020 02:44:08 +08:00


连接到:
Oracle Database 19c Enterprise Edition Release 19.0.0.0.0 - Production
Version 19.3.0.0.0
```

7.6 服务的启动和停止

如果通过网络(TCP/IP)访问数据库服务器,至少要启动监听器与数据库服务,即 Oracle neworcl TNSListener 和 OracleService**服务(**为数据库名)。如果本地即 Oracle 19c 客户端和服务端位于同一台计算机上(不是通过 TCP/IP 访问数据库服务器,至少要启动 OracleService＊＊服务(＊＊为数据库名)。

Oracle 服务的启动和关闭方式可以有如下 3 种方式。

(1) 在计算机的服务列表手动选择启动关闭服务。

(2) 使用操作命令,手动启动关闭服务。

(3) 在桌面上创建.bat 文件来启动和停止数据库系统。

① 创建启动文件：OrclStart.bat。

```
Net start OracleneworclTNSListener
Net start OracleServicesneworcl
```

② 创建停止文件：OrclStop.bat。

```
Net stop OracleneworclTNSListener
Net stop OracleServicesneworcl
```

7.7 项目案例

通过使用监听器命令来验证实例 neworl 的监听,其操作步骤如下：

步骤 1　启动监听,其代码如下：

```
LSNRCTL> start
启动 tnslsnr:请稍候...

TNSLSNR for 64-bit Windows: Version 19.0.0.0.0 - Production
系统参数文件为 D:\software\oracle19c\network\admin\listener.or
写入 D:\Oracle19c\diag\tnslsnr\LAPTOP-8BM7LBRI\listener\alert\log.xml 的日志信息
监听:(DESCRIPTION = (ADDRESS = (PROTOCOL = tcp)(HOST = 127.0.0.1)(PORT = 1521)))
监听:(DESCRIPTION = (ADDRESS = (PROTOCOL = ipc)(PIPENAME = \.\pipe\EXTPROC1521ipc)))

正在连接到 (DESCRIPTION = (ADDRESS = (PROTOCOL = TCP)(HOST = localhost)(PORT = 1521)))
LISTENER 的 STATUS
------------------------------------
别名                LISTENER
版本                TNSLSNR for 64-bit Windows: Version 19.0.0.0.0 - Production
启动日期            16-2月 -2020 13:46:23
正常运行时间        0 天 0 小时 0 分 16 秒
跟踪级别            off
安全性              ON: Local OS Authentication
SNMP                OFF
监听程序参数文件    D:\software\oracle19c\network\admin\listener.ora
监听程序日志文件 D:\Oracle19c\diag\tnslsnr\LAPTOP-8BM7LBRI\listener\alert\log.xml
监听端点概要...
```

```
    (DESCRIPTION = (ADDRESS = (PROTOCOL = tcp)(HOST = 127.0.0.1)(PORT = 1521)))
    (DESCRIPTION = (ADDRESS = (PROTOCOL = ipc)(PIPENAME = \.\pipe\EXTPROC1521ipc)))
(DESCRIPTION = (ADDRESS = (PROTOCOL = tcps)(HOST = LAPTOP-8BM7LBRI)(PORT = 5500))(Security =
(my_wallet_directory = D:\ORACLE19C\admin\neworcl\xdb_wallet))(Presentation = HTTP)(Session
= RAW))
服务摘要..
服务 "CLRExtProc" 包含 1 个实例.
    实例 "CLRExtProc", 状态 UNKNOWN, 包含此服务的 1 个处理程序...
服务 "neworcl" 包含 1 个实例.
    实例 "neworcl", 状态 READY, 包含此服务的 1 个处理程序...
服务 "neworclXDB" 包含 1 个实例.
    实例 "neworcl", 状态 READY, 包含此服务的 1 个处理程序...
命令执行成功
```

步骤 2 查看监听状态,其代码如下:

```
LSNRCTL> status
正在连接到 (DESCRIPTION = (ADDRESS = (PROTOCOL = TCP)(HOST = localhost)(PORT = 1521)))
LISTENER 的 STATUS
------------------------------------------------------------------------------
别名                    LISTENER
版本                    TNSLSNR for 64-bit Windows: Version 19.0.0.0.0 - Production
启动日期                16-2月 -2020 13:46:23
正常运行时间            0 天 0 小时 1 分 24 秒
跟踪级别                off
安全性                  ON: Local OS Authentication
SNMP                    OFF
监听程序参数文件        D:\software\oracle19c\network\admin\listener.ora
监听程序日志文件 D:\Oracle19c\diag\tnslsnr\LAPTOP-8BM7LBRI\listener\alert\log.xml
监听端点概要...
    (DESCRIPTION = (ADDRESS = (PROTOCOL = tcp)(HOST = 127.0.0.1)(PORT = 1521)))
    (DESCRIPTION = (ADDRESS = (PROTOCOL = ipc)(PIPENAME = \.\pipe\EXTPROC1521ipc)))
(DESCRIPTION = (ADDRESS = (PROTOCOL = tcps)(HOST = LAPTOP-8BM7LBRI)(PORT = 5500))(Security =
(my_wallet_directory = D:\ORACLE19C\admin\neworcl\xdb_wallet))(Presentation = HTTP)(Session
= RAW))
服务摘要..
服务 "CLRExtProc" 包含 1 个实例.
    实例 "CLRExtProc", 状态 UNKNOWN, 包含此服务的 1 个处理程序...
服务 "neworcl" 包含 1 个实例.
    实例 "neworcl", 状态 READY, 包含此服务的 1 个处理程序...
服务 "neworclXDB" 包含 1 个实例.
    实例 "neworcl", 状态 READY, 包含此服务的 1 个处理程序...
命令执行成功
```

步骤 3 查看监听服务，其代码如下：

```
LSNRCTL> service
正在连接到 (DESCRIPTION = (ADDRESS = (PROTOCOL = TCP)(HOST = localhost)(PORT = 1521)))
服务摘要..
服务 "CLRExtProc" 包含 1 个实例.
  实例 "CLRExtProc", 状态 UNKNOWN, 包含此服务的 1 个处理程序...
    处理程序:
      "DEDICATED" 已建立:0 已被拒绝:0
        LOCAL SERVER
服务 "neworcl" 包含 1 个实例.
  实例 "neworcl", 状态 READY, 包含此服务的 1 个处理程序...
    处理程序:
      "DEDICATED" 已建立:0 已拒绝:0 状态:ready
        LOCAL SERVER
服务 "neworclXDB" 包含 1 个实例.
  实例 "neworcl", 状态 READY, 包含此服务的 1 个处理程序...
    处理程序:
      "D000" 已建立:0 已被拒绝:0 当前: 0 最大: 1022 状态: ready
        DISPATCHER<machine: LAPTOP-8BM7LBRI, pid: 36328>
        (ADDRESS = (PROTOCOL = tcp)(HOST = LAPTOP-8BM7LBRI)(PORT = 64412))
命令执行成功
```

步骤 4 关闭监听，其代码如下：

```
LSNRCTL> stop
正在连接到 (DESCRIPTION = (ADDRESS = (PROTOCOL = TCP)(HOST = localhost)(PORT = 1521)))
命令执行成功
```

第8章 Oracle监控管理

PPT 视频讲解

如果 Oracle 在启动和运行过程中发生错误，可以通过查看诊断文件来确定发生错误的原因。在诊断文件中记载了日常数据库的活动信息，在实例的维护中，要经常查询诊断信息。

下面介绍自动诊断知识库（Automatic Diagnostic Repository，ADR），以及主要两种类型的诊断文件——告警日志文件与跟踪文件。

8.1 自动诊断知识库

1. 自动诊断知识库的概念

自动诊断知识库（ADR）是所有诊断信息的中心存储点。它包括各种转储文件、跟踪文件、日志和健康状况监视报表。所有实例（RDBMS 实例和 ASM 实例）都在 ADR 中创建自己的目录结构。

ADR 目录结构如图 8-1 所示。

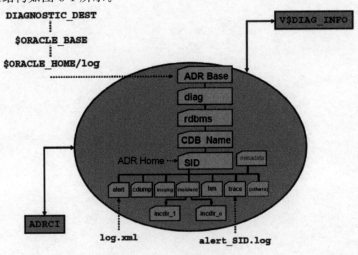

图 8-1　ADR 目录结构

关于 ADR 对应的目录位置可以查看 v$diag_info 视图,其代码如下:

```
SQL> col name format a30
SQL> col value format a50
SQL> set linesize 200
SQL> select * from v$diag_info;

   INST_ID NAME                 VALUE                                              CON_ID
---------- -------------------- -------------------------------------------------- ----------
         1 Diag Enabled         TRUE                                                    0
         1 ADR Base             D:\ORACLE19C                                            0
         1 ADR Home             D:\ORACLE19C\diag\rdbms\neworcl\neworcl                 0
         1 Diag Trace           D:\ORACLE19C\diag\rdbms\neworcl\neworcl\trace           0
         1 Diag Alert           D:\ORACLE19C\diag\rdbms\neworcl\neworcl\alert           0
         1 Diag Incident        D:\ORACLE19C\diag\rdbms\neworcl\neworcl\incident        0
         1 Diag Cdump           D:\Oracle19c\diag\rdbms\neworcl\neworcl\cdump           0
         1 Health Monitor       D:\ORACLE19C\diag\rdbms\neworcl\neworcl\hm              0
         1 Default Trace File   D:\ORACLE19C\diag\rdbms\neworcl\neworcl\trace\
                                 neworcl_ora_98788.trc                                  0

   INST_ID NAME                 VALUE                                              CON_ID
---------- -------------------- -------------------------------------------------- ----------
         1 Active Problem Count  1                                                      0
         1 Active Incident Count 2                                                      0
         1 ORACLE_HOME           D:\software\oracle19c                                  0

已选择 12 行.
```

其中,各子目录中包含的主要内容如下所述。

(1) Diag Alert:实例的告警日志,Diag Alert 对应的目录为 XML 格式的告警日志(对应为 log_x.xml)。

(2) Diag Cdump:核心文件。

(3) Diag Incident:每个意外事件的子目录中还包含该意外事件的所有跟踪转储。

(4) Diag Trace:跟踪文件(包括后台和用户转储目标)所在的位置。

(5) Diag Alert:对应的目录为文本格式的告警日志文件所在的目录。

(6) Default Trace File:是当前会话跟踪文件的名字。

2. 命令行工具(ADRCI)

ADR 命令行解释器(ADR Command Interpreter,ADRCI)是一个操作系统下的应用程序,可用于浏览所有 ADR 中的追踪文件,也可用于浏览 XML 格式的告警文件,能以命令行方式完成所有 OEM 支持工作台所允许的操作。

(1) ADRCI 的启动,其代码如下:

```
C:\WINDOWS\system32>ADRCI

ADRCI: Release 19.0.0.0.0 - Production on 星期日 2月 16 02:08:35 2020

Copyright (c) 1982, 2019, Oracle and/or its affiliates.  All rights reserved.

ADR base = "D:\Oracle19c"
ADRCI>
```

(2) ADRCI 简单操作

例如：列出 ADR 主目录的全路径，其代码如下：

```
ADRCI> show homes
ADR Homes:
diag\clients\user_lijie\host_1777851218_110
diag\clients\user_zlj\host_1777851218_110
diag\rdbms\neworcl\neworcl
diag\rdbms\testorcl\testorcl
diag\tnslsnr\LAPTOP-8BM7LBRI\listener
diag\tnslsnr\LAPTOP-8BM7LBRI\listener1

ADRCI> show alert

Choose the home from which to view the alert log:

1: diag\clients\user_lijie\host_1777851218_110
2: diag\clients\user_zlj\host_1777851218_110
3: diag\rdbms\neworcl\neworcl
4: diag\rdbms\testorcl\testorcl
5: diag\tnslsnr\LAPTOP-8BM7LBRI\listener
6: diag\tnslsnr\LAPTOP-8BM7LBRI\listener1
Q: to quit

Please select option:
```

3. 跟踪文件

跟踪文件中包含了大量而详细的诊断和调试信息。通过对跟踪文件的解读和分析，可以定位问题、分析问题和解决问题。

跟踪文件(Trace Files)可分为告警日志文件和跟踪文件两类。

(1) 告警日志文件(Alert Log File)：记录每天数据库的操作信息。

(2) 跟踪文件，跟踪文件又可以划分为以下两种。

① 后台进程跟踪文件(Background Process Trace Files)：记载后台进程等发生故障时

的重要信息。

② 用户跟踪文件(User Trace Files)：产生的用户跟踪信息，是由严重的用户错误引发的。

8.2 告警

告警日志文件是一类特殊的跟踪文件(Trace File)，其命名一般为 alert_<SID>.log。其中，SID 为 Oracle 数据库实例名称。数据库的告警日志主要按时间顺序记录消息和错误信息。

1. 告警日志的内容

告警日志的内容主要包括：

(1) 所有的内部错误(ORA-600)信息、块损坏错误(ORA-1578)信息，以及死锁错误(ORA-60)信息等。

(2) 管理操作，如 create、alter、drop 语句等，数据库启动、关闭以及日志归档的一些信息。

(3) 与共享服务器或调度进程相关功能的消息和错误信息。

(4) 物化视图的自动刷新过程中出现的错误。

(5) 动态参数的修改信息。

2. 告警日志的位置

(1) 正文格式的文件存放在每个 ADR 主目录的 TRACE 目录下。

(2) XML 格式的文件通过企业管理器和 ADRCI 命令行工具来浏览，并存放在每个 ADR 主目录中的 ALERT 目录下。

3. 告警日志监控功能

告警日志非常重要，必须保持告警日志的监控，应做到以下 4 点。

(1) 定期检查告警日志文件。

(2) 侦测系统内部错误(ORA-600)和有没有坏块(坏块指由于硬件或操作系统失误而导致的内容混乱的块)。

(3) 监控数据库的操作。

(4) 查看非默认的初始化参数。

8.3 跟踪

8.3.1 后台进程跟踪文件

后台进程跟踪文件就是由 DBWn、LGWR、SMON 等后台进程创建的后台跟踪文件。每个后台进程的错误都会产生后台进程跟踪文件。后台跟踪文件用于诊断和解决错误，当后台进程遇到错误就产生跟踪文件。

后台进程文件存放的位置是由 background_dump_dest 参数指定的目录，文件格式为

SID_dbwr_pid.trc、SID_smon_pid.trc 等。其中，pid 为创建文件的进程号，后台进程跟踪文件如图 8-2 所示。

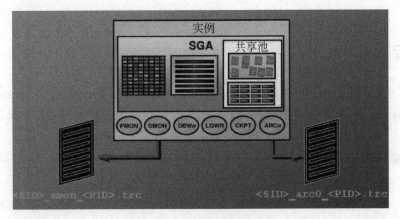

图 8-2　后台进程跟踪文件

可以采用如下命令，查看后台进程跟踪文件的具体存储路径。

```
SQL> show parameter background_dump_des

NAME                   TYPE      VALUE
--------------------   -------   -----------------------------------
background_dump_dest   string    D:\SOFTWARE\ORACLE19C\RDBMS\TRACE
```

8.3.2　用户进程跟踪文件

用户跟踪文件由连接到 Oracle 服务器的用户进程产生。这些文件仅在用户会话期间遇到错误时产生，用户可以通过执行 Oracle 跟踪事件来生成该类文件，所以文件中包含被跟踪的 SQL 语句的统计信息或用户的错误信息。服务器进程也可以产生此类文件。

1. 用户跟踪文件的位置

用户跟踪文件保存在 user_dump_dest 参数指定的目录中，在 Oracle 19c 中用户跟踪文件也存放在 TRACE 文件夹中，文件格式为 SID_ora_pid.trc。用户跟踪文件如图 8-3 所示。

2. 跟踪事件的设置

（1）查看当前数据是否开启追踪文件记录，其代码如下：

```
SQL> show parameter sql_trace;

NAME                             TYPE      VALUE
-------------------------------  --------  --------
sql_trace                        boolean   FALSE
```

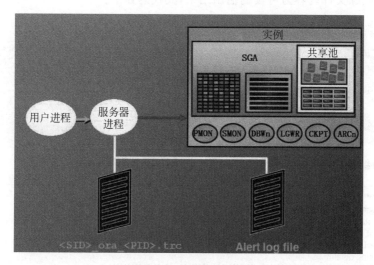

图 8-3　用户跟踪文件

（2）如果参数的值为 FALSE，就表示系统当前不会产生跟踪文件，采取如下操作可让系统产生跟踪文件。

```
SQL> alter session set sql_trace = true;

会话已更改.
```

（3）执行相应的 SQL 语句后，停止产生追踪文件的生成，其代码如下：

```
SQL> select * from test;

未选定行

SQL> alter session set sql_trace = false;

会话已更改.
```

（4）获取追踪文件存在的路径和文件名称，其代码如下：

```
SQL> col value format a50
SQL> set linesize 200
SQL> select * from v$diag_info where name = 'Default Trace File';

  INST_ID NAME                           VALUE                                                  CON_ID
---------- ------------------------------ -------------------------------------------------- ----------
        1 Default Trace File             D:\ORACLE19C\diag\rdbms\neworcl\neworcl\trace\neworcl_ora_110404.trc    0
```

(5) 转换跟踪文件内容为可读的输出结果,其代码如下:

```
C:\WINDOWS\system32>tkprof
D:\ORACLE19C\DIAG\RDBMS\NEWORCL\NEWORCL\TRACE\NEWORCL_ORA_110404.TRC   OUTPUT=D:\AA.TEXT

TKPROF: Release 19.0.0.0.0 - Development on 星期日 2月 16 16:27:52 2020

Copyright (c) 1982, 2019, Oracle and/or its affiliates.  All rights reserved.
```

转换后的结果如图 8-4 所示。

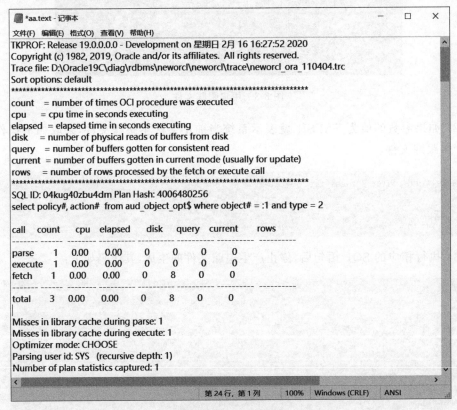

图 8-4 转换后的结果

在图 8-4 所示的结果中,各个分项关键字表示的含义如下所述。

① call:每次 SQL 语句的处理都分成 3 个步骤。
- parse:将 SQL 语句转换成执行计划,包括检查是否有正确的授权和所需要用到的表、列以及其他引用到的对象是否存在。
- execute:真正的 Oracle 执行语句。对于 insert、update、delete 操作,这步会修改数据;而对于 select 操作,这步骤就只是确定选择的记录。
- fetch:返回查询语句中所获得的记录。这步只有 select 语句会被执行。

② count:这个语句被 parse、execute、fetch 的次数。

③ cpu:这个语句对于所有的 parse、execute、fetch 所消耗的 CPU 的时间,以 s 为单位。

④ elapsed：这个语句所有消耗在 parse、execute、fetch 的总时间。

⑤ disk：从磁盘上的数据文件中物理读取的块的数量，一般来说更重要的是正在从缓存中读取的数据，而不是从磁盘上读取的数据。

⑥ query：在一致性读模式下，所有 parse、execute、fetch 所获得的 buffer 的数量。一致性模式的 buffer 是用于给一个长时间运行的事务提供一个一致性读的快照。

⑦ current：在 current 模式下所获得的 buffer 的数量。一般在 current 模式下执行 insert、update、delete 操作都会获取 buffer。如果在高速缓存区发现有新的缓存足够给当前的事务使用，那么这些 buffer 都会被读入缓存区中。

⑧ rows：所有 SQL 语句返回的记录数目。但是不包括子查询中返回的记录数目。对于 select 语句，返回记录是在 fetch 这一步骤；而对于 insert、update 和 delete 操作，返回记录是在 execute 这一步骤。

8.4 审计

8.4.1 审计概述

审计功能（Audit）用于监视用户所执行的数据库操作，并且 Oracle 会将审计跟踪结果存放在指定的地方。在 Oracle 12c 版本之后，推出了一套全新的审计架构，称为统一审计功能。新架构将现有审计跟踪统一为单一审计跟踪，从而简化了管理，提高了数据库生成的审计数据的安全性。

本书运行的数据库版本为 Oracle 19c。通过下面指令来观察与审计相关的参数。

```
SQL> show parameter audit;

NAME                                 TYPE        VALUE
------------------------------------ ----------- ------------------------------
audit_file_dest                      string      D:\ORACLE19C\ADMIN\NEWORCL\ADUMP
audit_sys_operations                 boolean     TRUE
audit_trail                          string      DB
unified_audit_sga_queue_size         integer     1048576
unified_audit_systemlog              boolean     FALSE
```

根据显示结果，下面详细讲解一下各主要参数的含义。

1. audit_trail 参数

该参数的值有以下 6 种情况。

（1）NONE：不开启。

（2）OS：启用数据库审计，并将数据库审计记录定向到操作系统文件，存储目录为 audit_file_dest。

（3）DB：开启审计功能，启用数据库审计，并将数据库的所有审计记录定向到数据库的 sys.aud$ 表目录，并由 audit_file_dest 指定审计文件存储目录。

(4) DB,EXTENDED：表示启用数据库审计,并将数据库所有审计记录定向到数据库的 SYS.AUD$表,另外填充 sys.aud$表的 sqlbind 列和 sqltext 列。

(5) XML：类似 OS,但是是将审计记录存放于 XML 格式的文件中。

(6) XML,EXTENDED：将审计记录存放于操作系统的 XML 文件中并填充 sql_text 和 sql_bind 信息。

注意：此参数无法动态修改,因此需要重启数据库才能生效,示例：

```
SQL>alter system set audit_trail = db scope = spfile;
```

对于 DB 和 DB,EXTENDED 两种审计模式,如果数据库是 read-only 模式的,那么数据库默认使用 OS 审计模式,无视已有设置(因为不能向数据库表写入数据)。

2. audit_file_dest 参数

该参数表示审计文件的位置。从结果中可以观察到,Oracle 19c 在 DB 模式下默认的审计文件位置为：D:\ORACLE19C\ADMIN\NEWORCL\ADUMP。

3. audit_sys_operations 参数

如果 audit_sys_operations 设置为 TRUE,那么用户 sys 的操作也会被审计。此参数控制以 sysdba 和 sysoper 权限登录的用户以及用户 sys 本身的登录,审计记录一定会写在操作系统文件中(无论 audit_trail 参数如何设置)。

8.4.2 统一审计

新引入的审计机制名为统一审计(Unified Audit),可以将其理解为一种更加轻便的精细审计。统一审计主要利用策略和条件,在 Oracle 数据库内部,有选择地执行有效的审计。新的基于策略的语法简化了数据库中的审计管理,并且能够基于条件加速审计。例如,审计策略可配置为根据特定 IP 地址、程序、时间段或连接类型(如代理身份验证)进行审计。此外,启用审计策略时,还可以轻松地将特定模式排除在审计之外。

可以使用如下命令观察 Oracle 19c 中统一审计的设置。

```
SQL> select parameter,value from v$option where parameter = 'Unified Auditing';

PARAMETER                                                        VALUE
---------------------------------------------------------------- --------
Unified Auditing                                                 FALSE
```

这种默认设置表示既支持那种标准的、精细的传统审计模式,也可以使用统一审计模式,这种默认的模式也被称作混合审计模式。为了需要,也可以自行启用统一模式。Oracle 19c 中的统一审计记录不在 aud$表中,而是存储在 audsys schema 下,可以通过 audsys.unified_audit_trail 表查询。在开启统一审计后,传统的 aud$和 dba_common_audit_trail 等不再包含任何审计记录。

混合模式审计和纯统一审计之间的差异如表 8-1 所示。

表 8-1 混合模式审计和纯统一审计之间的差异

模式	特征	启用
混合模式审核	既有传统审计,也有统一审计;可以使用统一审计工具和传统的审计工具	启用任何统一审计策略;无须重新启动数据库
统一审计	只有统一审计;只能使用统一审计工具	以 uniaud_on 选项链接二进制文件 Oracle,然后重新启动数据库

如果开启审计,那么传统审计模式就会被禁用,audit_trail 参数被忽略,aud$和 fga_log $等相关表不再新增任何审计记录,只能在 audsys.unified_audit_trail 中看到统一审计的记录。

8.4.3 审计策略

在 Oracle 11g 中,只要 audit_trail 的值不为 NONE,数据库就会默认审计一些操作。例如,logon、logout 和涉及数据库安全的操作等。而在 Oracle 12c 版本之后,系统也预设了一些统一审计策略,可以通过 audit_unified_enabled_policies 视图查看已生效的审计策略,通过 audit_unified_policies 可以查看所有预设审计策略,代码如下:

```
SQL> select policy_name, enabled_option, success, failure from
audit_unified_enabled_policies;

POLICY_NAME              ENABLED_OPTION   SUC  FAI
------------------------ ---------------- ---- -----
ORA_SECURECONFIG         BY USER          YES  YES
ORA_LOGON_FAILURES       BY USER          NO   YES
```

ora_logon_failures 在 Oracle 12.1.0.2 版本以后已从 ora_secureconfig 独立出来,并且只审计失败的登录。而从 audit_unified_policies 视图可以看出,ora_secureconfig 其实就包含了以前的那些默认标准审计项。混合模式下无论是传统审计记录,还是通过预设的统一审计策略进行的默认审计,其审计记录在 dba_common_audit_trail 中都可以查看。

下面介绍审计策略的常用操作。

1. 创建审计策略

创建定制化审计策略,语法如下:

```
create audit policy policy_name { {privilege_audit_clause [action_audit_clause] [role_audit_clause ]} | { action_audit_clause [role_audit_clause ] } | { role_audit_clause } } [when audit_condition evaluate per {statement|session|instance}] [container = {current|all}];
```

例如,创建一个审计策略,审计对表 hr.employees 的 select 操作语句如下:

```
SQL> create audit policy up1_2 actions select on hr.employees;
```

2. 开启关闭审计策略

```
audit policy<policy_name>;
noaudit policy<policy_name>;
```

3. 确认审计策略

```
SQL> select policy_name, audit_option_type, object_name, common from audit_unified_policies
where policy_name = upper('<policy_name>');
```

4. 查看有效审计策略

```
SQL> select user_name, policy_name, enabled_opt, success, failure from audit_unified_
enabled_policies;
```

5. 删除审计策略

```
SQL> drop audit policy<policy_name>;
```

在删除某审计策略之前，必须无效该审计策略，否则会报 ORA-46361 错误。

6. 查看审计结果

以前的审计功能，不同的组件会放在不同的位置存储。在统一审计功能下，存储和查看更加简单化，所有的审计结果都存放在新追加的 audsys schema 下，并可以通过字典表 unified_audit_trail 进行确认。

8.4.4 审计配置

1. 确认审计是否安装

```
SQL> select * from sys.aud$;
未选定行

SQL> select * from dba_audit_trail;
未选定行
```

如果当进行上述查询时，发现表不存在，则说明审计相关的表还没有安装。注意，安装结束后，需要重启数据库。代码如下：

```
SQL> @D:\SOFTWARE\ORACLE19C\RDBMS\ADMIN\CATAUDIT.SQL
视图已创建.
同义词已创建.
注释已创建.
注释已创建.
```

```
授权成功.
授权成功.
...

会话已更改.
```

2. 审计表空间迁移

审计表默认安装在表空间 SYSTEM。为了方便管理,也可以将相关表迁移到其他表空间。代码如下:

```
SQL>alter table aud$ move tablespace new_tablespace;
SQL>alter index I_aud1 rebuild online tablespace new_tablespace;
SQL>alter table audit$ move tablespace new_tablespace;
SQL>alter index i_audit rebuild online tablespace new_tablespace;
SQL>alter table audit_actions move tablespace new_tablespace;
SQL>alter index i_audit_actions rebuild online tablespace new_tablespace;
```

3. 审计级别

当开启审计功能后,可在以下 3 个级别对数据库进行审计。

(1) Statement:按语句来审计,比如 audit table 会审计数据库中所有的 create table、drop table、truncate table 语句。

(2) Privilege:按权限来审计,若用户使用了该权限,则被审计。如执行 grant select any table to a,在执行了 audit select any table 语句后,当用户 a 访问了用户 b 的表时(如 select * from b.t)会用到 select any table 权限,故会被审计。注意,用户是自己表的所有者,所以用户访问自己的表不会被审计。

(3) Object:按对象审计,只审计 on 关键字指定对象的相关操作,如 audit alter、delete、drop、insert on employees by hr;这里会对用户 hr 的 employees 表进行审计,但同时使用了 by 子句,所以只会对用户 hr 发起的操作进行审计。

提示:Oracle 没有提供对 schema 中所有对象的审计功能,只能一个一个地对象审计,对于后面创建的对象,Oracle 则提供 on default 子句来实现自动审计,如执行 audit drop on default by access 后,对于随后创建的对象的 drop 操作都会被审计。

4. 其他审计选项

(1) by access:每一个被审计的操作都会生成一条 audit trail。

(2) by session:一个会话里面同类型的操作只会生成一条 audit trail,默认为 by session。

(3) whenever successful:只有操作成功(dba_audit_trail 中 returncode 字段为 0)才会被审计。

(4) whenever not successful:反之如果省略该子句,不管操作成功与否,都会被审计。

5. 审计相关视图

(1) dba_audit_trail：保存所有的 audit trail，实际上它只是一个基于 aud$的视图。其他的视图 dba_audit_session、dba_audit_object、dba_audit_statement 都只是 dba_audit_trail 的一个子集。

(2) dba_stmt_audit_opts：可以用来查看 statement 审计级别的审计选项，即数据库设置过哪些 statement 级别的审计。

(3) all_def_audit_opts：用来查看数据库用 on default 子句设置了哪些默认对象审计。

(4) audit_unified_policies：记录所有统一审计策略。

(5) audit_unified_enabled_policies：记录所有已启用的统一审计策略。

(6) unified_audit_trail：记录所有统一审计结果。

8.4.5 审计的使用

针对传统审计还需要了解常用的关于审计的命令。

1. 开启审计

```
alter system set audit_sys_operations = true scope = spfile;
alter system set audit_trail = db,extended scope = spfile;
```

2. 关闭审计

```
alter system set audit_trail = none;
```

3. 撤销审计

```
noaudit all on 'object_name';
```

4. 清除审计

```
delete from sys.aud$;
delete from sys.aud$ where oj$name = 'object_name';
truncate table sys.aud$;
```

8.5 项目案例

视频讲解

8.5.1 审计策略实例

创建一个针对用户 hr 中 employees 表查询的特定审计策略。具体操作步骤如下所述。

步骤 1　创建审计策略，其代码如下：

```
SQL> create audit policy up1 actions select on hr.employees;
```

审计策略已创建.

步骤 2 启用审计策略,其代码如下:

```
SQL> audit policy up1;
```

审计已成功.

步骤 3 确定审计策略是否创建成功,其代码如下:

```
SQL> set linesize 200
SQL> col policy_name format a20
SQL> col object_name format a15
SQL> select policy_name,audit_option_type,object_name,common
  2  from audit_unified_policies where policy_name = upper('up1');

POLICY_NAME          AUDIT_OPTION_TYPE  OBJECT_NAME     COM
-------------------- ------------------ --------------- ----
UP1                  OBJECT ACTION      EMPLOYEES
```

步骤 4 查询目前有效的审计策略,其代码如下:

```
SQL> col policy_name format a20
SQL> col user_name format a20
SQL> select policy_name,enabled_option,success,failure  from
audit_unified_enabled_policies;

POLICY_NAME          ENABLED_OPTION   SUC FAI
-------------------- ---------------- --- ---
ORA_SECURECONFIG     BY USER          YES YES
ORA_LOGON_FAILURES   BY USER          YES YES
UP1                  BY USER          YES YES
```

步骤 5 禁用审计策略,其代码如下:

```
SQL> noauditx policy up1;
```

审计未成功.

步骤 6 删除审计策略,其代码如下:

```
SQL> drop audit policy up1;
```

审计策略已删除.

步骤 7 查询目前有效的审计策略,其代码如下:

```
SQL> col policy_name format a20
SQL> col user_name format a20
SQL> select policy_name,enabled_option,success,failure  from
audit_unified_enabled_policies;

POLICY_NAME          ENABLED_OPTION    SUC  FAI
-------------------- ---------------   ---  ---
ORA_SECURECONFIG     BY USER           YES  YES
ORA_LOGON_FAILURES   BY USER           YES  YES
```

8.5.2 语句审计实例

视频讲解

步骤 1 设置审计,其代码如下:

```
SQL> show user;
USER 为 "SYS"
SQL> audit table by hr by access;
```

审计已成功.

步骤 2 查看审计语句级设置,其代码如下:

```
SQL> col audit_option for a10
SQL> col user_name for a10
SQL> select user_name,audit_option,success,failure from dba_stmt_audit_opts;

USER_NAME  AUDIT_OPTI  SUCCESS    FAILURE
---------  ----------  ---------  ---------
HR         TABLE       BY ACCESS  BY ACCESS
```

步骤 3 语句审计实例。
(1) 登录用户 hr,创建 course 表,其代码如下:

```
SQL> conn hr
输入口令:
已连接.
SQL> create table course
```

```
2  (课程名称 varchar(22),
3  学分 varchar(4),
4  班级 varchar2(20),
5  大外教师 varchar2(10));
```

表已创建.

(2) 在 course 表上创建唯一索引 coucno, 其代码如下:

```
SQL> create unique index coucno on course(课程名称);
```

索引已创建.

(3) 查看审计结果, 其代码如下:

```
SQL> col owner for a30
SQL> col onj_name for a20
SQL> select username,to_char(timestamp,'yyyy-mm-dd hh24:mi:ss'),
  2  owner,action_name,obj_name from dba_audit_object;

USERNAME    TO_CHAR(TIMESTAMP,'   OWNER ACTION_NAME    OBJ_NAME
------------------------------------------------------------
HR2020-02-16 02:31:15 HR              CREATE TABLE COURSE
```

(4) 停止审计, 其代码如下:

```
SQL> noaudit table by hr;
审计未成功.

SQL> commit;
提交完成.
```

(5) 删除审计, 其代码如下:

```
SQL> conn sys as sysdb
输入口令:
已连接.
SQL> delete from sys.aud$;

已删除 1 行.
```

8.5.3 权限审计实例

步骤 1 创建临时用户 u1, 并为其分配权限, 其代码如下:

```
SQL> create user u1 identified by u1;
用户已创建.

SQL> grant create session,create table to u1;
授权成功.
```

步骤 2　设置审计,其代码如下:

```
SQL> audit delete table , insert table by u1 by access whenever successful;

审计已成功.
```

步骤 3　查看权限级设置,其代码如下:

```
SQL> select user_name,audit_option,success,failure from dba_stmt_audit_opts;

USER_NAME   AUDIT_OPTI SUCCESS    FAILURE
---------   ---------- --------   --------
U1          INSERT TAB BY ACCESS  NOT SETLE
U1          DELETE TAB BY ACCESS  NOT SETLE
```

步骤 4　权限审计实例。
(1) 以用户 u1 登录,创建表 test 并插入数据,其代码如下:

```
SQL> conn u1
输入口令:
已连接.
SQL> create table test(a int);

表已创建.

SQL> insert into test values(1);
insert into test values(1)
            *
第 1 行出现错误:
ORA - 01950: 对表空间 'USERS' 无权限

SQL> commit;

提交完成.

SQL> select * from test;
```

未选定行

SQL> delete from test;

已删除 0 行.

(2) 查看审计结果,其代码如下:

```
SQL> select username,to_char(timestamp,'yyyy-mm-dd hh24:mi:ss'),
  2  owner,action_name,obj_name from dba_audit_object;

USERNAME            TO_CHAR(TIMESTAMP,'OWNER    ACTION_NAME    OBJ_NAME
--------            --------------------------  -----------    --------
U1                  2020-02-16 02:44:25 U1       DELETE         TEST
```

(3) 停止权限审计,其代码如下:

```
SQL> noaudit delete table,insert table by u1 whenever successful;

审计未成功.
```

8.5.4 对象审计实例

视频讲解

步骤1 对象审计设置,其代码如下:

```
SQL> audit select on u1.test by access whenever successful;

审计已成功.
```

步骤2 查询对象审计设置,其代码如下:

```
SQL> select owner,object_name,object_type  from dba_obj_audit_opts where owner='U1';

OWNER              OBJECT_NAME                       OBJECT_TYPE
---------------    ----------------------------      --------------------
U1                 TEST                              TABLE
```

步骤3 对象审计实例。
(1) 以用户 u1 登录,向表 test 中插入数据,其代码如下:

```
SQL> conn u1
输入口令:
已连接.
SQL> insert into test values(1);
```

```
insert into test values(1)
            *
第 1 行出现错误:
ORA-01950: 对表空间 'USERS' 无权限

SQL> select * from test;
未选定行

SQL> commit;
提交完成.
```

(2) 查询审计结果,其代码如下:

```
SQL>  select username,to_char(timestamp,'yyyy-mm-dd hh24:mi:ss'),
  2   owner,action_name,obj_name from dba_audit_object;

USERNAME   TO_CHAR(TIMESTAMP,          OWNER   ACTION_NAME     OBJ_NAME
--------   ------------------------    -----   -------------   --------
U1         2020-02-16 02:51:27         U1      SELECT          TEST
U1         2020-02-16 02:44:25         U1      DELETE          TEST
```

(3) 停止对象审计,其代码如下:

```
SQL> noaudit select,insert,delete,delete on u1.test whenever successful;
审计未成功.
```

第9章

数据库的归档模式管理

PPT 视频讲解

在数据库的使用过程中,可能出现断电、死机等意外情况。当出现意外时,如何保证数据的有效性、一致性和完整性呢?Oracle 作为大型关系数据库管理系统,必须要通过合理的机制确保在任何情况下都不会出现数据丢失,通过合理的配置重做日志可以实现并完成这项任务。重做日志文件可以分为以下两种。

(1) 在线重做日志(Online Redo Log Files):又称为联机重做日志,指 Oracle 以 SQL 脚本的形式实时记录数据库的数据更新。

(2) 归档重做日志(Archive Redo Log Files):简称为归档日志,指当条件满足时,Oracle 将在线重做日志以文件形式保存到硬盘。

在第 3 章已经详细讲解了在线重做日志的原理以及具体操作,本章重点讲解何为数据库的归档模式,如何将数据库设置为归档模式以及如何管理归档重做日志。

9.1 非归档模式与归档模式

Oracle 数据库可运行在两种模式下:归档模式和非归档模式。作为一个合格的 DBA,应当深入了解这两种日志操作模式的特点,并且保证数据库在合适的日志操作模式下运行。

非归档模式与归档模式如图 9-1 所示。

1. 非归档模式的利与弊

当 Oracle 数据库运行在非归档模式下,则意味着数据库发生日志切换后,重做日志文件不需要归档。非归档模式下,数据库只能提供实例级别的故障恢复,如果需要介质恢复的时候 Oracle 就爱莫能助了。此时,Oracle 只能把数据库恢复到过去的某个时间点上,前提是在这个时间点完全冷备份了数据库。但从备份时间点到故障发生期间的所有数据都丢失了。而且当 Oracle 运行在非归档模式下时,数据库不能提供在线的表空间备份,换言之热备份是不可用的。

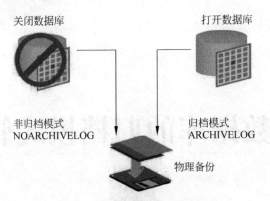

图 9-1 非归档模式与归档模式

非归档模式的重做日志历史如图 9-2 所示。

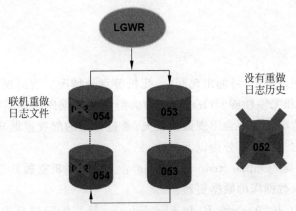

图 9-2 非归档模式的重做日志历史

2. 归档模式的利与弊

当 Oracle 数据库运行在归档模式下,控制文件确定了发生日志切换后的重做日志文件在归档前是不能被日志写进程重用的。如果启动了归档模式,后台归档进程就会启动,将当前重做日志中的记录存储到归档目录中,形成归档日志文件。Oracle 数据库能从实例介质的故障中得到恢复,即最近一次的数据库全备及备份以来的所有归档日志的备份都将得到恢复。

归档模式可以分为手动归档和自动归档。顾名思义,手动归档需要 DBA 的干预;而在自动归档模式下,Oracle 会自己完成归档任务。

归档模式下的重做日志文件历史如图 9-3 所示。

概括来说,重做日志归档模式与非归档模式的最重要的区别就是当前的重做日志切换以后会不会被归档进程复制到归档目的地。但是具体选择哪种模式,则需要从实际情况出发综合考虑。对于生产环境,一般选用归档模式。下面将介绍当前模式的查询以及模式间的切换。

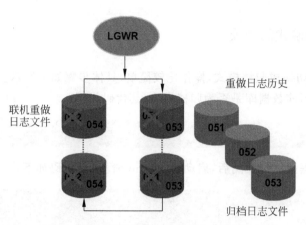

图 9-3 归档模式下的重做日志文件历史

9.2 归档模式设置

9.2.1 归档模式的查询

归档模式的查询方法有以下两种。
（1）使用如下语句查询归档状态，其代码如下：

```
SQL> select name,log_mode from v$database;

NAME       LOG_MODE
--------   ----------
NEWORCL    NOARCHIVELOG
```

（2）采用以下语句查看数据库归档模式，其代码如下：

```
SQL> archive log list;
数据库日志模式              非存档模式
自动存档                   禁用
存档终点                   D:\software\oracle19c\RDBMS
最早的联机日志序列           4
当前日志序列                6
```

在默认情况下，Oracle 数据库采用的是非归档模式。

通过以上查询结果得知：数据库处于非归档模式下，自动归档功能被禁用，最早的重做日志序列号是 4，当前日志序列号是 6。

9.2.2 归档模式间切换

将非归档模式切换成归档模式,操作比较简单,具体步骤如下所示。

步骤 1　查看当前数据库是否为归档模式,其代码如下:

```
SQL> archive log list;
```

步骤 2　将数据库正常关闭后,启动到 Mount 阶段,其代码如下:

```
SQL> shutdown immediate
SQL> startup mount
```

步骤 3　将数据库设定为归档模式,其代码如下:

```
SQL> alter database archivelog;
```

步骤 4　将数据库打开供联机使用,其代码如下:

```
SQL> alter database open;
```

步骤 5　再次查看数据库运行模式,确认是否成功更改为归档模式并备份数据库,其代码如下:

```
SQL> archive log list;
```

由非归档模式切换至归档模式的步骤示意如图 9-4 所示。

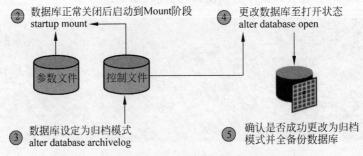

图 9-4　非归档模式切换至归档模式的步骤示意

9.2.3 归档模式设置

当数据库处于归档模式时,在进行了日志切换后必须要归档重做日志。因为日志组只有在归档后才能被覆盖,所以如果不归档该日志组,当下次切换到该日志组时就会导致 LGWR 处于等待状态。为了避免出现这种情况,必须要及时归档重做日志,归档重做日志有自动归档和手工归档两种方法。

1. 自动归档

若出现以下 3 种情形,则 DBA 可以选择在实例启动前或实例启动后启用自动归档功能。

(1) 如果要在实例启动前启用自动归档功能,需要将初始化参数文件中的参数 log_archive_start 的值改为 true。操作步骤如下所示。

步骤 1 以管理员身份登录数据库,其代码如下:

```
SQL> conn / as sysdb
```

步骤 2 修改服务器端初始化参数,其代码如下:

```
SQL> alter system set log_archive_start = true scope = spfile;
```

步骤 3 重新启动数据库,使参数生效,其代码如下:

```
SQL> shutdown;
SQL> startup;
```

步骤 4 在进行以上的自动归档设置之后,可以通过查询确定是否自动归档,其代码如下:

```
SQL> show parameter log_archive_start
```

(2) 如果实例启动时没启用自动归档,不必先关闭实例和对初始化参数进行修改,就可以直接使用如下语句启用自动归档功能。

```
SQL> alter system archive log start;
```

执行上述语句后,无须重新启动实例,即可立即启用自动归档功能,但是再次启动实例时,自动归档功能仍然关闭。

(3) 关闭自动归档,其代码如下:

```
SQL> alter system archive log stop;
```

2. 手工归档

一旦禁用自动归档功能,DBA 必须手动执行命令,且对所有或特定的联机重做日志文件都需进行手动归档。在归档模式下,如果禁用自动归档模式,但是又没有及时进行手工归档,LGWR 进程无法写入下一个已经写满的可用重做日志组,这时数据库将被暂时停止执行,直到完成对重做日志组的归档为止。

对日志文件进行归档的语法为：

```
alter system archive log current | all
```

归档后，可通过 v$archived_log 视图确认，查询代码如下：

```
SQL> select recid, name, first_time from v$archived_log;
```

9.3 归档重做日志

9.3.1 归档日志切换

通过上述操作，数据库的运行模式已经切换为归档模式了。通过存档终点参数可以追溯到归档日志的路径，代码如下：

```
SQL> show parameter db_recovery_file_dest;

NAME                                 TYPE        VALUE
------------------------------------ ----------- ------------------------------
db_recovery_file_dest                string
db_recovery_file_dest_size           big integer 0
```

但是从显示结果发现，db_recovery_file_dest 的值为空，也就是代表该路径下内容为空。这是为什么呢？为了再一次确认，再次运行以下语句。

```
SQL> select destination,binding,target,status from v$archive_dest;

DESTINATION                        BINDING    TARGET    STATUS
---------------------------------- ---------- --------- ----------
D:\software\oracle19c\rdbms\       MANDATORY  PRIMARY   VALID
                                   OPTIONAL   PRIMARY   INACTIVE
                                   OPTIONAL   PRIMARY   INACTIVE
                                   OPTIONAL   PRIMARY   INACTIVE
                                   OPTIONAL   PRIMARY   INACTIVE
.......
已选择 31 行
```

查询 v$archive_dest 数据字典视图可获知：在存档终点目录下的归档日志文件是强制的，这就代表之前的设置应该没有问题，那么问题出在什么地方呢？为什么归档日志文件不存在呢？

这是因为在前一段时间没有任何 DML 操作，所以重做日志文件不可能被填满，也就不可能产生重做日志的切换，当然也就不可能产生归档日志文件。为了产生归档日志文件，可

以在使用重做日志的切换命令,其代码如下:

```
SQL> alter system switch logfile;

系统已更改.
```

再次查询 v$archived_log,结果显示归档日志成功生成。

```
SQL> select name from v$archived_log;

NAME
--------------------------------------------------
D:\SOFTWARE\ORACLE19C\RDBMS\ARC0000000008_1022597296.0001
```

9.3.2 归档进程

由上述可知,重做日志写进程(LGWR)负责将重做日志缓冲区的数据顺序地写到重做日志文件中;而归档进程(ARCn)是把切换后的重做日志文件复制到归档日志文件。也就是说,LGWR 是读内存写外存(硬盘);而 ARCn 是读外存写外存。一般来说,内存的读取速度为外存的 100～100000 倍。

LGWR 和 ARCn 具体的操作过程示意如图 9-5 所示。

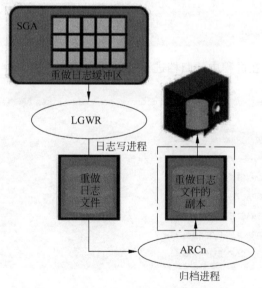

图 9-5　LGWR 和 ARCn 具体的操作过程示意

LGWR 是将内存中的数据写入重做日志文件,是内存读磁盘写;而 ARCn 是将重做日志文件写入归档文件,是磁盘读磁盘写。显然 LGWR 的读写效率或者读写速度比 ARCn 的快,而在频繁发生 DML 操作的数据库中,可能发生由于归档慢而重做日志写入速度快所造成的数据库被暂时停止执行的情况,此时数据库就是等待 ARCn 将当前的重做日志数据写入归档文件。

为了匹配二者的速度,可以考虑修改归档进程参数(log_archive_max_processes),它的作用就是提高归档进程的数量,以启动更多的归档进程完成写归档的过程。具体操作如下。

步骤 1　查询当前归档进程,其代码如下:

```
SQL> show parameter log_archive_max_processes;

NAME                                 TYPE        VALUE
------------------------------------ ----------- --------
log_archive_max_processes            integer     4
```

步骤 2　更改归档进程个数,其代码如下:

```
SQL> alter system set log_archive_max_processes = 5;

系统已更改.
```

步骤 3　查询调整后参数,其代码如下:

```
SQL> show parameter log_archive_max_processes;

NAME                                 TYPE        VALUE
------------------------------------ ----------- --------
log_archive_max_processes            integer     5
```

步骤 4　查询目前在进程中的归档进程,其代码如下:

```
SQL> select pid,pname,program from v$process where pname like 'ARC%';

   PID PNAME PROGRAM
------ ----- ----------------------------
    36 ARC0  ORACLE.EXE (ARC0)
    38 ARC1  ORACLE.EXE (ARC1)
    39 ARC2  ORACLE.EXE (ARC2)
    40 ARC3  ORACLE.EXE (ARC3)
    42 ARC4  ORACLE.EXE (ARC4)
```

9.3.3　归档日志文件和归档目录

通过修改参数 log_archive_max_processes 解决了重做日志 LGWR 写入速度和 ARCn 写入速度的不匹配问题。接下来要讨论归档日志的安全问题。

1. 查询归档进程文件物理路径的状态

首先可以利用参数 log_archive_dest_state 观察并查询所有归档日志文件物理路径的状态。

```
SQL> show parameter log_archive_dest_state

NAME                                 TYPE        VALUE
------------------------------------ ----------- ----------
log_archive_dest_state_1             string      enable
log_archive_dest_state_10            string      enable
log_archive_dest_state_11            string      enable
log_archive_dest_state_12            string      enable
log_archive_dest_state_13            string      enable
log_archive_dest_state_14            string      enable
log_archive_dest_state_15            string      enable
......

已选择 31 行。
```

从结果中可以看到，所有的归档日志共 31 个，所有路径的状态都是 enable，也就是都是可用的。若想观察每个具体归档文件的最终归档路径，也可以从数据字典 v$archive_dest 中获得。

2. 查询归档的目标归档路径

通过下列语句可以查询归档的目标归档路径，代码如下：

```
SQL> select destination,binding,target,status from v$archive_dest;

DESTINATION                      BINDING      TARGET          STATUS
-------------------------------  -----------  --------------  --------
D:\software\oracle19c\rdbms\     MANDATORY    PRIMARY         VALID
                                 OPTIONAL     PRIMARY         INACTIVE
...

已选择 31 行。
```

上述视图结果中，binding 代表归档路径强制要求。当值为 OPTIONAL 时，表示若该目录没有成功归档，也可以切换日志；当值为 MANDATORY 时，表示强制归档成功。status 代表状态，当值为 VALID 时，代表可用，而 INACTIVE 则代表未活动。

通过以上查询结果可获知，目前只有一个归档日志文件的物理路径是强制性的，其他 30 个文件的模式值都为可选。这也就是说，Oracle 系统在任何时刻只能保证一组归档日志文件是好的。实际上此时的数据库运行在十分脆弱的状态下,因为如果这组唯一没有问题的归档日志文件再损坏的话,将使数据库的全部恢复成为一件不可能的事。

3. 查询成功归档日志文件组的最低数

如何避免数据库运行在这样一种十分脆弱的状态下呢?为此,Oracle 引入了另外一个动态参数——log_archive_min_succeed_dest,通过定义这一参数的值来限定 Oracle 系统必须保证成功的归档日志文件组数(最低要求),其代码如下:

```
SQL> show parameter log_archive_min_succeed_dest;

NAME                                 TYPE        VALUE
------------------------------------ ----------- --------
log_archive_min_succeed_dest         integer     1
```

目前能够保证成功归档日志文件组数为 1。下面利用语句更新该参数值,以调整成功组数。

```
SQL> alter system set log_archive_min_succeed_dest = 2;
alter system set log_archive_min_succeed_dest = 2
                 *
第 1 行出现错误:
ORA-02097: 无法修改参数,因为指定的值无效
ORA-16020: 可用的目标少于由 LOG_ARCHIVE_MIN_SUCCEED_DEST 指定的数量
```

更新语句执行发现错误,也就是说如果强行将其修改为 2,就会报错。可以通过更改参数,设置存储路径,从而增加最低成功归档文件数。具体操作步骤以及源码参考本章 9.5 节的项目案例。

数据库开启归档模式之后,要时刻关注归档目录的空间使用情况。设置合理的归档删除策略。如果归档空间不足,就会导致无法切换日志,出现 ORA-6014 错误,提示:log n sequence # not archived,no available destinations。虽然开启归档模式会增加系统 I/O 压力,但是在生产系统中还是建议将数据库置于归档模式。

9.4 归档模式下相关数据视图和脚本

1. 归档模式下相关的数据视图

(1) v$database:视图中的 log_mode 代表数据库归档模式。

(2) v$archived_log:控制文件中已经归档的日志文件信息。

(3) v$archive_dest:所有归档目标。

(4) v$archive_processes:已启动的归档进程状态。

2. 查看归档日志的脚本量

(1) 估算系统每秒产生的日志量和每天产生的日志量的脚本,其代码如下:

```
SQL> set serveroutput on
SQL> declare
  2     ac              number;
  3     sec_redo        number;
  4     day_redo        number;
  5     sec_redo_90 number;
  6     day_redo_90 number;
  7     str             varchar(100);
  8   begin
  9     select count(*) into ac from v$database where log_mode = 'NOARCHIVELOG';
 10     if ac = 1
 11     then
 12       dbms_output.put_line('The database is running on NOARCHIVELOG mode,No archivelog!');
 13     else
 14       str := '';
 15       for i in (select destination
 16                 from v$archive_dest
 17                 where status = 'VALID'
 18                 and destination is not null) loop
 19         str := str || ',' || i.destination;
 20       end loop;
 21       dbms_output.put_line('Archive dest is ' || substr(str, 2));
 22       select trunc(sum((blocks * block_size) / 1024) /
 23                    ((max(first_time) - min(first_time)) * 24 * 3600)),
 24              trunc(sum((blocks * block_size) / 1024 / 1024) /
 25                    ((max(first_time) - min(first_time))))
 26         into sec_redo, day_redo
 27         from v$archived_log;
 28       dbms_output.put_line('每秒产生日志的频率:' || sec_redo || '(KB)');
 29       dbms_output.put_line('每天产生日志的频率:' || day_redo || '(MB)');
 30       dbms_output.put_line('最近3个月统计数据:');
 31       select trunc(sum((blocks * block_size) / 1024) /
 32                    ((max(first_time) - min(first_time)) * 24 * 3600)),
 33              trunc(sum((blocks * block_size) / 1024 / 1024) /
 34                    ((max(first_time) - min(first_time))))
 35         into sec_redo_90, day_redo_90
```

```
36      from v$archived_log
37      where first_time>SYSDATE - 91;
38      dbms_output.put_line('每秒产生日志的频率：'|| sec_redo_90 || '(KB)');
39      dbms_output.put_line('每天产生日志的频率:'|| day_redo_90 || '(MB)');
40   end if;
41  end;
42  /
Archive dest is D:\software\oracle19c\rdbms、d:\backup\
每秒产生日志的频率: 0 (KB)
每天产生日志的频率:62(MB)
最近3个月统计数据:
每秒产生日志的频率: 0 (KB)
每天产生日志的频率:62(MB)

PL/SQL 过程已成功完成.
```

(2) 观察归档日志的每日生成量，其代码如下：

```
SQL> select lpad(to_char(first_time, 'yyyymmdd'), 12) "Date",
  2        trunc(sum(blocks * block_size) / 1024 / 1024) "size(MB)",
  3        count( * )"count"
  4    from v$archived_log
  5   where first_time>SYSDATE - 31
  6   group by lpad(to_char(first_time, 'yyyymmdd'), 12)
  7   order by 1;

Date                   size(MB)     COUNT
------------------     --------     -----
20200202                   685          4
20200206                   171          2
20200209                   373          4
20200212                   337          2
20200215                   468          8
```

(3) 通过以下脚本可以观察归档日志的每小时生成量，其代码如下：

```
SQL> select lpad(to_char(first_time, 'hh24'), 4) "Hour",    trunc(sum(blocks * block_size)
/ 1024 / 1024) "size(MB)",      count( * )"count"   FROM v$archived_log  where first_time>
sysdate - 31 group by lpad(to_char(first_time, 'hh24'), 4) order by 1;

Hour       size(MB)     COUNT
-------    --------     -----
02              171         2
11              685         4
13               90         2
```

```
17              24          4
20             354          2
21               3          2
22             369          2
23             337          2
```

已选择 8 行.

9.5 项目案例

9.5.1 归档模式切换实例

视频讲解

先查看数据库日志操作模式，然后将数据库由非归档模式切换为归档模式。

步骤 1　查看当前数据库是否为归档模式，其代码如下：

```
SQL> archive log list;
数据库日志模式                  非存档模式
自动存档                        禁用
存档终点                        D:\software\oracle19c\RDBMS
最早的联机日志序列              4
当前日志序列                    6
```

步骤 2　关闭数据库，然后装载数据库。

因为只有在数据库处于 Mount 状态下时才能修改日志操作模式，所以必须先关闭数据库，再装载数据库。但是需要注意，关闭数据库时不能使用 shutdown abort 命令。

```
SQL> conn / as sysdba;
已连接.
SQL> shutdown;
数据库已经关闭.
已经卸载数据库.
Oracle 例程已经关闭.
SQL> startup mount;
Oracle 例程已经启动.
Total System Global Area 34675092 bytes
Fixed Size 453012 bytes
Variable Size 29360128 bytes
Database Buffers 4194304 bytes
Redo Buffers 667648 bytes
数据库装载完毕.
```

步骤 3　修改归档模式，其代码如下：

```
SQL> alter database archivelog;
```

数据库已更改.

步骤 4　打开数据库,其代码如下:

```
SQL> alter database open;
```

数据库已更改.

在步骤 4 中要注意:在修改了日志操作模式之后,必须重新备份数据库。

步骤 5　查看当前归档模式,其代码如下:

```
SQL> archive log list;
数据库日志模式              存档模式
自动存档                    启用
存档终点                    d:\backup\
最早的联机日志序列           4
下一个存档日志序列           6
当前日志序列                 6
```

9.5.2　归档日志文件实例

为了能保证添加归档日志强制存储路径并保证最低要求,下面将介绍具体的解决办法。

步骤 1　查找有关归档日志的参数信息,其代码如下:

```
SQL> select destination,binding,target,status from v$archive_dest;

DESTINATION                       BINDING      TARGET          STATUS
------------------------------    ----------   -------------   ----------
D:\software\oracle19c\rdbms\      MANDATORY    PRIMARY         VALID
                                  OPTIONAL     PRIMARY         INACTIVE
...

已选择 31 行.
```

通过数据字典视图显示发现,Oracle 准备了可用的 31 个可以进行归档日志存储的归档目录。目前所有状态都是可用的,但是只有一个归档路径非空,其他 30 个均为空,代表没有设置具体的归档目录。

步骤 2　更改参数,设置存储路径,其代码如下:

```
SQL> alter system set
log_archive_dest_1 = 'location = D:\SOFTWARE\ORACLE19C\RDBMS\ MANDATORY';
```

系统已更改
```
SQL> alter system set log_archive_dest_2 = 'location = D:\BACKUP\ mandatory';
```
系统已更改.

步骤 3 更改后,多个归档目录路径会同时保存归档日志,其代码如下:

```
SQL> show parameter log_archive_dest_;

NAME                              TYPE         VALUE
--------------------------------- ------------ ------------------------------------
log_archive_dest_1                string       location = D:\SOFTWARE\ORACLE19C\RDBMS\ mandatory
log_archive_dest_2                string       location = d:\BACKUP\ mandatory
log_archive_dest_state_1          string       enable
log_archive_dest_state_2          string       enable
......
```

步骤 4 更改参数,保证系统写到物理路径下的归档日志文件至少两个是成功的,其代码如下:

```
SQL> alter system set log_archive_min_succeed_dest = 2;
```

系统已更改.

步骤 5 查询当前系统保证成功的归档日志组数,其代码如下:

```
SQL> show parameter log_archive_min_succeed_dest;

NAME                              TYPE         VALUE
--------------------------------- ------------ ------------------------------------
log_archive_min_succeed_dest      integer      2
```

步骤 6 查询参数调整后的归档日志文件信息,其代码如下:

```
SQL> show parameter log_archive_min_succeed_dest;

NAME                              TYPE         VALUE
--------------------------------- ------------ ------------------------------------
log_archive_min_succeed_dest      integer      2
```

步骤 7 查询参数调整后的归档日志文件信息,其代码如下:

```
SQL> select destination,binding,target,status from v$archive_dest;

DESTINATION                       BINDING      TARGET              STATUS
--------------------------------- ------------ ------------------- -------
```

```
D:\SOFTWARE\ORACLE19C\RDBMS\    MANDATORY  PRIMARY         VALID
D:\BACKUP\                      MANDATORY  PRIMARY         VALID...
```

已选择 31 行。

步骤 8　切换用重做日志，其代码如下：

```
SQL> alter system switch logfile;
系统已更改。

SQL> alter system switch logfile;
系统已更改。
```

步骤 9　通过下面的语句可观察到两个归档目录下都进行了归档日志的存取。

```
SQL> select name from v$archived_log;

NAME
--------------------------------------------------
D:\SOFTWARE\ORACLE19C\RDBMS\ARC0000000007_1031954683.0001
D:\BACKUP\ARC0000000007_1031954683.0001
D:\SOFTWARE\ORACLE19C\RDBMS\ARC0000000008_1031954683.0001
D:\BACKUP\ARC0000000008_1031954683.0001
```

步骤 10　关闭控制归档路径。

使用动态初始化参数 log_archive_dest_state_n 来关闭归档目的地，其代码如下：

```
SQL> alter system set log_archive_dest_state_2 = defer;
alter system set log_archive_dest_state_2 = defer
*
第 1 行出现错误：
ORA - 02097：无法修改参数，因为指定的值无效
ORA - 16028：新 LOG_ARCHIVE_DEST_STATE_2 导致少于 LOG_ARCHIVE_MIN_SUCCEED_DEST
所需的目的地数量
```

关闭归档路径的同时要考虑最低归档成功数。如果低于要求，则拒绝关闭归档状态。

第 3 部分

数据库运维

第 10 章　Oracle 备份

第 11 章　Oracle 恢复

第 12 章　Oracle 数据的移动

第 13 章　Oracle 闪回技术

第 14 章　Oracle 并发与一致性

第 15 章　Oracle 优化

第10章 Oracle备份

PPT 视频讲解

10.1 数据库故障

现实工作中有很多情况都可能造成数据丢失,造成数据丢失的主要因素可能是介质故障,如磁盘损坏、磁头碰撞、瞬时强磁场干扰;用户的错误操作;服务器的彻底崩溃;计算机病毒或者不可预料的因素,如自然灾害、电源故障、盗窃等都会造成数据库的多种故障,且解决每种故障的办法也不尽相同。下面列举一下 Oracle 中的几种常见故障以及解决办法。

1. 语句错误的典型问题与解决办法

语句错误的典型问题与解决办法如表 10-1 所示。

表 10-1　语句错误的典型问题与解决办法

典型问题	解决方法
尝试在表中输入无效的数据	与用户合作来验证并更正数据
尝试在权限不足时执行操作	提供适当的对象或系统权限
尝试分配未成功分配的空间	启用可恢复的空间分配; 增加所有者限额; 添加表空间的空间
应用程序中的逻辑错误	与开发人员合作来更正程序错误

2. 用户进程错误的典型问题与解决办法

用户进程错误的典型问题与解决办法如表 10-2 所示。

表 10-2　用户进程错误的典型问题与解决办法

典型问题	解决方法
用户执行了异常断开连接操作	通常不需要 DBA 的操作就可以解决用户进程错误。实例后台进程会回退未提交的更改并解除锁定
用户会话已异常终止	
用户遇到了终止会话的程序错误	

3. 网络故障的典型问题与解决办法

网络故障的典型问题与解决办法如表 10-3 所示。

表 10-3　网络故障的典型问题与解决办法

典型问题	解决方法
监听程序失败	配置备份监听程序和连接时故障转移
网络接口故障	配置多个网卡
网络连接失败	配置备份网络连接

4. 用户错误的典型问题与解决办法

用户错误的典型问题与解决办法如表 10-4 所示。

表 10-4　用户错误的典型问题与解决办法

典型问题	解决方法
用户无意中删除或修改了数据	回退或使用闪回查询进行恢复
用户删除了表	从回收站恢复表

5. 实例错误的典型问题与解决办法

实例错误的典型问题与解决办法如表 10-5 所示。

表 10-5　实例错误的典型问题与解决办法

典型问题	解决方法
断电	使用 startup 命令重新启动实例。从实例错误中恢复时是自动执行的
硬件故障	
后台进程出现错误	
紧急关闭命令	

6. 介质故障的典型问题与解决办法

介质故障的典型问题与解决办法如表 10-6 所示。

表 10-5　介质故障的典型问题与解决办法

典型问题	解决方法
硬盘驱动故障	1. 从备份中还原受影响的文件；
磁盘控制器故障	2. 通知数据关于新文件的位置；
删除或损坏了数据库文件	3. 通知应用重做信息来恢复文件

10.2 数据库备份概述

数据库备份是 DBA 十分重要的一项任务,使用备份的数据库文件可以在数据库出现人为或设备故障时迅速地恢复数据,保证数据库系统对外提供持续、一致的数据库服务。

备份是数据库的一个副本,具体内容包括数据文件、控制文件等,也可以是逻辑备份。通过备份,DBA 可以有效地防止不可预测的数据丢失或应用程序错误造成的数据丢失。

从不同的角度来看待备份,它有很多种分类方法。

(1) 从物理与逻辑的角度来分类。

物理备份与逻辑备份两种。

① 物理备份:对数据库操作系统的物理文件(数据文件、控制文件和日志文件)的备份。物理备份又可以分为脱机备份(冷备份)和联机备份(热备份)。其中,脱机备份是在关闭数据库的时候进行的;联机备份是以归档日志的方式对运行的数据库进行的备份。也可以使用 Oracle 的恢复管理器(RMAN)或操作系统命令来进行数据库的物理备份。

② 逻辑备份:对数据库逻辑组件(如表和存储过程等数据库对象)的备份。逻辑备份的手段很多,如传统的 EXP、数据泵(EXPDP)、数据库闪回技术等第三方工具,都可以进行数据库的逻辑备份。

(2) 从数据库是否联机的角度划分。

冷备份与热备份两种。

① 冷备份:也称为脱机备份,是将关键性文件复制到另外的位置的一种说法。冷备份发生在数据库已经正常关闭的情况下,当正常关闭时会提供给一个完整的数据库。对于备份而言,冷备份是最快和最安全的方法。

② 热备份:也称为联机备份,是指数据库在打开的状态进行的备份。此时用户可以继续访问数据库,执行 DML 操作。但是要求数据库必须运行在归档模式下,只有在归档模式下才可以对整个数据库、单独的表以及数据文件进行备份。

(3) 从数据库的备份角度分类。

完全备份、增量备份和差异备份三种。

① 完全备份:每次对数据库进行完整备份。当发生数据丢失的情况时,完全备份无需依赖其他信息即可实现 100% 的数据恢复,其恢复时间最短且操作最方便。

② 增量备份:只需备份自上次完全备份或增量备份后被修改的文件即可。优点是备份数据量小,需要的时间短;缺点是恢复的时候需要依赖以前的备份记录,风险较大。

③ 差异备份:自从上次完全备份之后被修改过的文件的备份。从差异备份中恢复数据的时间较短,因此只需要两份数据,即最后一次完整备份和最后一次差异备份,缺点是每次备份需要的时间较长。

10.2.1 冷备份

当 DBA 正常关闭 Oracle 服务器时(通常以 shutdown normal 或 shutdown immediate 指令关闭),系统会自动触发一次检查点事件。在数据库关闭之前,此检查点首先将缓存中的数据写入数据文件与联机日志文件,再把当前的检查点序号写入所有数据文件、联机日志文件和控制文件中。所以,正常关闭后的数据库是一个状态一致的数据库,即它的所有文件的时间点是一致的。此时,将数据库的所有数据文件、联机日志文件和控制文件复制出来形成的备份称为全库脱机备份,该备份可直接用于整个数据库的恢复(但只能恢复到备份时间点)。

冷备份主要备份的是数据库所有的数据文件、控制文件和日志文件。由于其他文件都比较小,所以为了简化备份的过程,通常在脱机备份时可将其他文件一起备份。

下面介绍具体冷备份的主要操作步骤。

步骤1　启动 SQL Plus,以 sysdba 身份登录数据库。
查询当前数据库所有数据文件、控制文件、联机重做日志文件、初始化参数文件的位置。

步骤2　正常关闭数据库。

步骤3　复制所有数据文件、控制文件、联机重做日志文件以及初始化参数文件等其他配置文件到备份磁盘。

可以直接在操作系统中使用复制、粘贴方式进行,也可以在 SQL Plus 环境中使用下列形式的操作系统命令完成。

```
SQL> host copy 原文件名称    目标路径名称
```

步骤4　重新启动数据库。

进行冷备份的操作步骤示意如图 10-1 所示。具体的操作步骤与源码参见 10.5.1 节的冷备份实例。

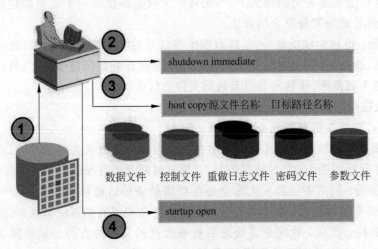

图 10-1　冷备份的操作步骤示意

冷备份具有如下优点。
(1) 只需复制文件即可,是非常快速的备份方法。
(2) 只需将文件再复制回去,就可以恢复到某一时间点上。
(3) 与数据库归档的模式相结合可以使数据库很好地恢复。
(4) 维护量较少,但安全性却相对较高。

10.2.2 热备份

虽然冷备份具有简单、快捷等优点,但是在很多情况下,如数据库运行于 24×7 状态时(每天工作 24 小时,每周工作 7 天),如果没有足够的时间关闭数据库进行冷备份,就只能采用热备份。

热备份又可称为联机备份。热备份是指对数据库运行在归档模式下进行的数据文件、控制文件、归档日志文件等的备份。

在 SQL Plus 环境中进行数据库完全热备份的操作步骤如下所述。

步骤 1　启动 SQL Plus,以 sysdba 身份登录数据库。

步骤 2　将数据库设置为归档模式。

由于热备份是数据库处于归档模式下的备份,因此在热备份之前,可以执行 archive log list 命令,以查看当前数据库是否处于归档日志模式。如果没有处于归档日志模式,就先将数据库转换为归档模式,并启动自动存档。

步骤 3　以表空间为单位,进行数据文件备份。

(1) 查看当前数据库有哪些表空间以及每个表空间中有哪些数据库文件,其代码如下:

```
SQL> SELECT tablespace_name,file_name from dba_data_files order by tablespace_name;
```

(2) 分别对每个表空间中的数据文件进行备份。

① 将需要备份的表空间(如 USERS)设置为备份状态。

```
SQL> alter tablespace users begin backup;
```

② 将表空间中所有的数据文件复制到备份磁盘。

```
SQL> host copy
host xcopy D:\ORACLE19C\ORADATA\NEWORCL\USERS01.DBF
D:\BACKUP_NEWORCL\HOTBACK;
```

③ 结束表空间的备份状态。

```
SQL> alter tablespace users end backup;
```

对数据库中所有表空间分别采用上述步骤①～④进行备份。

步骤 4　备份控制文件。通常应该在数据库物理结构做出修改之后,如添加、删除或重命名数据文件,添加、删除或修改表空间,添加或删除重做日志文件和重做日志文件组等,都需要重新备份控制文件。其有以下两种方法:

(1) 创建二进制映像文件。

```
SQL> alter database backup controlfile to 'E:\APP\ADMIN\BACKUP\CONTROL.BAK';
```

(2) 创建一个正文追踪文件。

```
SQL> alter database backup controlfile to trace;
```

步骤5　备份其他物理文件。
(1) 归档当前的联机重做日志文件。

```
SQL> alter system archive log current;
```

归档当前的联机重做日志文件,也可以通过日志切换完成。

```
SQL> alter system switch logfile;
```

(2) 备份归档重做日志文件,将所有的归档重做日志文件复制到备份磁盘中。
(3) 备份初始化参数文件,将初始化参数文件复制到备份磁盘中。

10.3　项目案例

视频讲解

10.3.1　冷备份实例

下面以具体实例来介绍冷备份的操作步骤。
步骤1　启动 SQL Plus,以 sysdba 身份登录数据库。登录后查询当前数据库所有数据文件、控制文件、联机重做日志文件等的位置。
(1) 查询数据文件,其代码如下:

```
SQL> select file_name from dba_data_files;

FILE_NAME
-------------------------------------------------
D:\ORACLE19C\ORADATA\NEWORCL\SYSTEM01.DBF
D:\ORACLE19C\ORADATA\NEWORCL\SYSAUX01.DBF
D:\ORACLE19C\ORADATA\NEWORCL\USERS01.DBF
D:\ORACLE19C\ORADATA\NEWORCL\STUDENT01.DBF
D:\ORACLE19C\ORADATA\NEWORCL\UNDOTBS01.DBF
```

(2) 查询控制文件,其代码如下:

```
SQL> select name from v$controlfile;

NAME
-------------------------------------------------
D:\ORACLE19C\ORADATA\NEWORCL\CONTROL01.CTL
D:\ORACLE19C\ORADATA\NEWORCL\CONTROL02.CTL
D:\BACKUP\CONTROL03.CTL
```

(3) 查询联机重做日志文件，其代码如下：

```
SQL> select group#,member,from v$logfile;

   GROUP#  MEMBER
---------- ---------------------------------------------
        2  D:\ORACLE19C\ORADATA\NEWORCL\REDO02-2.LOG
        2  D:\ORACLE19C\ORADATA\NEWORCL\REDO02.LOG
        1  D:\ORACLE19C\ORADATA\NEWORCL\REDO01.LOG
        3  D:\ORACLE19C\ORADATA\NEWORCL\REDO03-1.LOG
        3  D:\ORACLE19C\ORADATA\NEWORCL\REDO03-2.LOG
        3  D:\ORACLE19C\ORADATA\NEWORCL\REDO03-3.LOG

已选择 6 行.
```

(4) 查询参数文件，其代码如下：

```
SQL> show parameter pfile

NAME      TYPE     VALUE
--------- -------- -------------------------------------------
Spfile    string   D:\SOFTWARE\ORACLE19C\DATABASE\SPFILENEWORCL.ORA
```

当然，参数文件也可以不用备份，其不影响冷备份效果。

步骤 2　正常方式关闭数据库。

```
SQL> shutdown immediate
数据库已经关闭.
已经卸载数据库.
ORACLE 例程已经关闭.
```

步骤 3　复制所有数据文件、重做日志文件以及控制文件并粘贴到备份磁盘。

可以直接在操作系统中使用复制、粘贴方式进行备份，也可以使用下面的操作系统命令完成。

```
SQL> host xcopy 原文件名称　目标路径名称
```

将命令设置为批处理文件，可以使备份操作变得简单，如 coolBak.sql 文件即为一个自动化实例。

```
SQL> host xcopy D:\ORACLE19C\ORADATA\NEWORCL\*.*   D:\BACKUP_NEWORCL\20200217
D:\Oracle19c\oradata\NEWORCL\CONTROL01.CTL
D:\Oracle19c\oradata\NEWORCL\CONTROL02.CTL
D:\Oracle19c\oradata\NEWORCL\REDO01.LOG
```

```
D:\Oracle19c\oradata\NEWORCL\REDO02-2.LOG
D:\Oracle19c\oradata\NEWORCL\REDO02.LOG
D:\Oracle19c\oradata\NEWORCL\REDO03-1.LOG
D:\Oracle19c\oradata\NEWORCL\REDO03-2.LOG
D:\Oracle19c\oradata\NEWORCL\REDO03-3.LOG
D:\Oracle19c\oradata\NEWORCL\REDO03.LOG
D:\Oracle19c\oradata\NEWORCL\STUDENT01.DBF
D:\Oracle19c\oradata\NEWORCL\SYSAUX01.DBF
D:\Oracle19c\oradata\NEWORCL\SYSTEM01.DBF
D:\Oracle19c\oradata\NEWORCL\TEMP01.DBF
D:\Oracle19c\oradata\NEWORCL\UNDOTBS01.DBF
D:\Oracle19c\oradata\NEWORCL\USERS01.DBF
复制了 15 个文件
```

步骤 4 重新启动数据库，其代码如下：

```
SQL> startup
Oracle 例程已经启动.

Total System Global Area 2550136752 bytes
Fixed Size                  9031600 bytes
Variable Size             570425344 bytes
Database Buffers         1962934272 bytes
Redo Buffers                7745536 bytes
数据库装载完毕.
数据库已经打开.
```

视频讲解

10.3.2 热备份实例

步骤 1 启动 SQL Plus，以身份 sysdba 登录数据库。

```
请输入用户名:   sys as sysdb
输入口令:
已连接到空闲例程.
```

步骤 2 确定将数据库是否运行在归档模式，其代码如下：

```
SQL> archive log list;
数据库日志模式            存档模式
自动存档                  启用
存档终点                  d:\backup\
最早的联机日志序列         7
下一个存档日志序列         9
当前日志序列               9
```

在热备份之前，先将日志进行一次切换，并将在线日志存储到归档日志中，其代码如下：

```
SQL> after system switch logfile
数据库日志模式              存档模式
自动存档                    启用
存档终点                    d:\backup\
最早的联机日志序列          7
下一个存档日志序列          9
当前日志序列                9
```

再次确认表空间的信息，以便所有表空间都能实现热备份。

```
SQL> select tablespace_name,file_name from dba_data_files order by tablespace_name;

TABLESPACE_NAME                FILE_NAME
------------------------------ --------------------------------------
STUDENT                        D:\ORACLE19C\ORADATA\NEWORCL\STUDENT01.DBF
SYSAUX                         D:\ORACLE19C\ORADATA\NEWORCL\SYSAUX01.DBF
SYSTEM                         D:\ORACLE19C\ORADATA\NEWORCL\SYSTEM01.DBF
UNDOTBS1                       D:\ORACLE19C\ORADATA\NEWORCL\UNDOTBS01.DBF
USERS                          D:\ORACLE19C\ORADATA\NEWORCL\USERS01.DBF
```

步骤 3　进行数据文件备份。

下面以表空间 SYSTEM 为例，进行 system01.dbf 文件的备份。在备份的过程中，为了提高效率，建议将代码提前准备好，以便一次性输入，其代码如下：

```
SQL> alter tablespace system begin backup;
表空间已更改.

SQL> host xcopy   D:\ORACLE19C\ORADATA\NEWORCL\SYSTEM01.DBF
D:\BACKUP_NEWORCL\HOTBACK;
D:\ORACLE19C\ORADATA\NEWORCL\SYSTEM01.DBF
复制了 1 个文件

SQL> alter tablespace system end backup;
表空间已更改.
```

接下来依次进行表空间 UNTO、SYSAUX、USER 与 STUDENT 的数据文件备份，其代码如下：

```
SQL> alter tablespace undotbs1 begin backup;
表空间已更改.

SQL> host xcopy   D:\ORACLE19C\ORADATA\NEWORCL\UNDOTBS01.DBF   D:\BACKUP_NEWORCL\HOTBACK;
```

```
D:\ORACLE19C\ORADATA\NEWORCL\UNDOTBS01.DBF
复制了 1 个文件.

SQL> alter tablespace undotbs1 end backup;
表空间已更改.

SQL> alter tablespace sysaux begin backup;
表空间已更改.

SQL> host xcopy  D:\ORACLE19C\ORADATA\NEWORCL\SYSAUX01.DBF   D:\BACKUP_NEWORCL\HOTBACK;
D:\ORACLE19C\ORADATA\NEWORCL\SYSAUX01.DBF
复制了 1 个文件.

SQL> alter tablespace sysaux end backup;
表空间已更改.

SQL> alter tablespace users begin backup;
表空间已更改.

SQL> host xcopy D:\ORACLE19C\ORADATA\NEWORCL\USERS01.DBF   D:\BACKUP_NEWORCL\HOTBACK;
D:\ORACLE19C\ORADATA\NEWORCL\USERS01.DBF
复制了 1 个文件.

SQL> alter tablespace users end backup;
表空间已更改.

SQL> alter tablespace student begin backup;
表空间已更改.

SQL> host xcopy D:\ORACLE19C\ORADATA\NEWORCL\STUDENT01.DBF   D:\BACKUP_NEWORCL\HOTBACK;
D:\ORACLE19C\ORADATA\NEWORCL\STUDENT01.DBF
复制了 1 个文件.

SQL> alter tablespace student end backup;
表空间已更改.
```

步骤 4　备份控制文件,其代码如下:

```
SQL> alter database backup controlfile to
'D:\Backup_neworcl\hotBack;

数据库已更改.


SQL> alter database backup controlfile to trace;
数据库已更改.
otBack\CONTROL.BKP';

数据库已更改.
```

步骤5 备份其他物理文件。
(1) 将重做日志文件归档,其代码如下:

```
SQL> alter system archive log current;

系统已更改.
```

(2) 备份归档重做日志文件和初始化参数文件。
① 备份归档重做日志文件:将所有的归档重做日志文件复制到备份磁盘中。
② 备份初始化参数文件:将初始化参数文件复制到备份磁盘中。
步骤6 热备份结束后,查看动态新能字典中记录的备份信息,其代码如下:

```
SQL> select * from v$backup;

     FILE#  STATUS              CHANGE#    TIME              CON_ID
---------- ------------------ ---------- ----------------- ----------
         1  NOT ACTIVE          3387170    17-2月 -20             0
         2  NOT ACTIVE          3387243    17-2月 -20             0
         3  NOT ACTIVE          3387207    17-2月 -20             0
         4  NOT ACTIVE          3387190    17-2月 -20             0
         7  NOT ACTIVE          3387226    17-2月 -20             0
```

10.3.3 冷备份自动化实例

为了方便执行备份,同时也为了在备份的过程中减少异常发生,可以将冷备份和热备份进行自动化设置。
首先介绍冷备份自动化实例的具体步骤。
步骤1 D盘下创建待备份的文件夹 BackUpNeworcl。
步骤2 编辑冷备份脚本 CoolBak.sql,如图 10-2 所示。

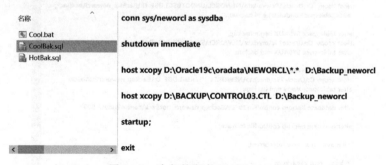

图 10-2 冷备份脚本 CoolBak.sql

步骤3 编辑批命令文本 Cool.bat,如图 10-3 所示。
步骤4 创建冷备份的快捷方式,如图 10-4 所示。

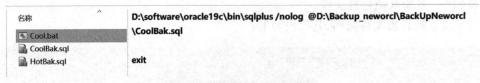

图 10-3　批命令文本 Cool.bat

图 10-4　冷备份的快捷方式

在创建了冷备份的快捷方式之后,为了数据库的安全,可以双击冷备份的快捷方式图标,将冷备份进行定时设置,以使数据库进行定时备份。

10.3.4　热备份自动化实例

热备份自动化实例的具体操作步骤如下所述。

步骤1　在 D 盘下创建待备份的文件夹 BackUp_Neworcl。

步骤2　创建热备份脚本 HotBak.sql,放置在 D 盘,如图 10-5 所示。

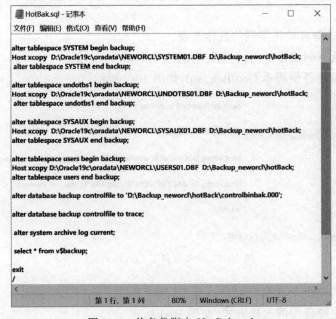

图 10-5　热备份脚本 HotBak.sql

步骤 3　创建调用热备份的批处理文件 Hot.bat，如图 10-6 所示。

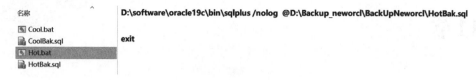

图 10-6　热备份的批处理文件 Hot.bat

步骤 4　创建热备份的快捷方式，更改图标，效果如图 10-7 所示。

图 10-7　热备份的快捷方式图标

双击热备份的快捷方式，可设置热备份的自动运行时间，以保证热备份的定时执行。

第11章

Oracle恢复

PPT 视频讲解

11.1 数据库恢复概述

数据库恢复就是指在计算机系统发生故障后,可利用已备份的数据文件或控制文件,重新建立一个完整的数据库。根据不同的情况,恢复也有以下 4 种分类方法。

1. 实例恢复与介质恢复

(1) 实例恢复:当 Oracle 实例出现失败后,Oracle 自动进行的恢复。

(2) 介质恢复:当存放数据库的介质出现故障时所做的恢复。

2. 完全恢复与不完全恢复

(1) 完全恢复:将数据库恢复到数据库失败时的状态。这种恢复是通过装载数据库备份并应用全部的重做日志来实现的。

(2) 不完全恢复:将数据库恢复到数据库失败前的某一时刻的状态。这种恢复是通过装载数据库备份并应用部分的重做日志来实现的。在进行不完全恢复后,必须在启动数据库时使用 resetlogs 选项重设联机重做日志。

3. 物理恢复与逻辑恢复

逻辑恢复是指利用 Oracle 提供的导入工具将逻辑备份形成的转储文件导入数据库内部,进行数据库的逻辑恢复;否则为物理恢复。

4. 归档模式下的热恢复和非归档模式下的冷恢复

利用在非归档模式下的备份文件恢复即非归档下的冷备份恢复。而归档模式下的热恢复需要先应用归档日志,当数据文件和控制文件一致之后才能打开数据库,如果归档日志和在线日志没有损坏,就可以完全恢复数据库,也就是说可以不丢失数据。

11.1.1 实例恢复

实例恢复过程是由系统的 SMON 自动进行的。当数据库正在运行时,系统突然发生意外,此时,由于数据的更新是先在内存中更新,然后在一定的时机再写入数据文件,所以有以下两种情况产生。

(1) 某些事物已经提交,但是在内存中没来得及存入到数据文件,这些事务应该被重做。

(2) 某些事务较长,部分数据已经写入数据文件,但是整个事物还没有提交,这些事务应该撤销。

当重新启动实例后,系统会自动进行实例恢复,其操作步骤示意如图 11-1 所示。

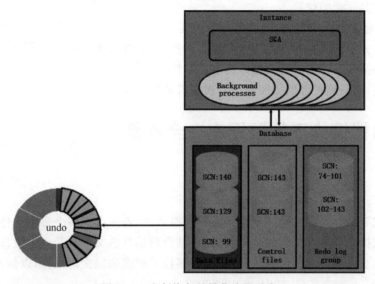

图 11-1 实例恢复的操作步骤示意

实例恢复的阶段通常可以概括为以下步骤。
步骤 1 启动数据库(数据文件不同步)。
步骤 2 前滚。
步骤 3 文件中包含提交和未提交数据。此时数据库已经打开,允许用户连接。
步骤 4 数据库打开。
步骤 5 回滚。
步骤 6 数据文件中仅包含提交数据。

11.1.2 介质恢复

介质恢复和实例恢复的机制是类似的。所不同的是,介质恢复是当存储的数据文件出现故障的时候进行的,并且介质恢复无法自动进行,必须手动输入 recover database 或者 recover datafile 命令来实施。一般来说,介质恢复是从一个恢复的数据文件为起点进行恢复,因此在做介质恢复的时候,需要使用归档日志。

11.2 非归档模式下的数据库恢复

非归档模式下的数据库恢复主要是利用在非归档模式下的冷备份恢复数据库,把最近时间点全库脱机备份全部复制过来,即用备份的数据文件、联机日志文件、控制文件覆盖并替换数据库当前的所有数据文件、联机日志文件和控制文件。如果有数据文件没有被替换,就说明它的时间点与其他文件的时间点不一致,数据库不能打开使用。

非归档模式下的数据库恢复操作步骤相对简单,如下所述。

步骤1 关闭数据库。

步骤2 把上一次脱机全库备份的所有文件(数据文件、联机日志文件和控制文件)全部一起复制到其原来的位置。

步骤3 启动数据库系统,完成数据库恢复。

非归档模式下的数据库恢复方式具有过程简单、操作方便的特点。但该方式要求数据库在能关闭的情况下才能进行数据库的备份和恢复,因此不太适用。另外,这种方式的恢复是不完全恢复,它只能使数据库回到上一次备份数据的时间点。

11.3 归档模式下的数据库完全恢复

11.3.1 概念

归档模式下的数据库完全恢复是指在归档模式下,若一个或多个数据文件损坏,则可利用热备份的数据文件替换损坏的数据文件,再结合归档日志文件和联机重做日志文件,采用前滚技术重做自备份以来的所有改动,采用回滚技术回滚未提交的操作,以恢复到数据库故障时刻的状态。

11.3.2 恢复的级别

(1) 数据库级完全恢复:主要应用于所有或多数数据文件损坏的恢复。其操作步骤如下:

步骤1 如果数据库没有关闭,则强制关闭数据库。

步骤2 利用备份的数据文件还原所有损坏的数据文件。

步骤3 将数据库启动到 Mount 状态。

步骤4 执行数据库恢复命令。

步骤5 打开数据库。

(2) 表空间级完全恢复:对指定表空间中的数据文件进行恢复。

以表空间 STUDENT 的数据文件 student01.dbf 被损坏为例来模拟表空间级的完全恢复。

① 数据库处于装载状态下的恢复。其操作步骤如下:

步骤1 如果数据库没有关闭,则强制关闭数据库。

```
SQL> shutdown abort
```

步骤 2　利用备份的数据文件 student01.dbf 还原被损坏的数据文件 student01.dbf。
步骤 3　将数据库启动到 Mount 状态。

```
SQL> startup mount
```

步骤 4　执行表空间恢复命令。

```
SQL> recover tablespace student
```

步骤 5　打开数据库。

```
SQL> alter database open
```

② 数据库处于打开状态下的恢复,其操作步骤如下:
步骤 1　如果数据库已经关闭,就将数据库启动到 Mount 状态。

```
SQL> startup mount
```

步骤 2　将损坏的数据文件设置为脱机状态。

```
SQL> alter database datafile '...\STUDENT01.DBF' offline
```

步骤 3　打开数据库。

```
SQL> alter database open
```

步骤 4　将损坏的数据文件所在的表空间脱机。

```
SQL> alter tablespace student offline for recover
```

步骤 5　利用备份的数据文件 student01.dbf 还原损坏的数据文件 student01.dbf。
步骤 6　执行表空间恢复命令。

```
SQL> recover tablespace student
```

步骤 7　将表空间联机。

```
SQL> alter tablespace student online
```

如果数据文件被损坏时数据库正处于打开状态,就可以直接执行步骤 4～步骤 7 的操作。
(3) 数据文件级完全恢复:针对特定的数据文件进行恢复。
以下以数据文件 student01.dbf 被损坏为例来模拟数据文件级的完全恢复,具体操作步骤如下所示。

① 数据库处于装载状态下的恢复,其操作步骤如下:
步骤 1　如果数据库没有关闭,就强制关闭数据库。

```
SQL> shutdown abort
```

步骤 2　利用备份的数据文件...\STUDENT01.DBF 还原损坏的数据文件 student01.dbf。
步骤 3　将数据库启动到 Mount 状态。

```
SQL> startup mount
```

步骤 4　执行数据文件恢复命令。

```
SQL> recover datafile '...\STUDENT01.DBF'
```

步骤 5　将数据文件联机。

```
SQL> alter database datafile
D:\STUDENT.DBF' online
```

步骤 6　打开数据库。

```
SQL> alter database open;
```

② 数据库处于打开状态下的恢复,其操作步骤如下:
步骤 1　如果数据库已经关闭,则将数据库启动到 Mount 状态。

```
SQL> startup mount
```

步骤 2　将损坏的数据文件设置为脱机状态。

```
SQL> alter database datafile '...\STUDENT01.DBF' offline;
```

步骤 3　打开数据库。

```
SQL> alter database open;
```

步骤 4　利用备份的数据文件 example01.dbf 还原损坏的数据文件 student01.dbf。
步骤 5　执行数据文件恢复命令。

```
SQL> recover datafile '...\STUDENT01.DBF';
```

步骤 6　将数据文件联机。

```
SQL> alter database datafile '...\STUDENT01.DBF' online;
```

步骤 7　如果数据文件损坏时数据库正处于打开状态,则可以直接执行步骤 2、步骤 4~

步骤 6 即可。

注意：数据库级的完全恢复只能在数据库装载但没有打开的状态下进行，而表空间级完全恢复和数据文件级完全恢复可以在数据库处于装载状态或打开的状态下进行。

11.3.3　归档模式下数据库完全恢复

归档模式下数据库完全恢复的代码如下：

```
recover [automatic] [from 'location']
[database|tablespace tspname
|datafile dfname]
```

说明：
automatic：进行自动恢复，不需要 DBA 提供重做日志文件名称。
location：指定归档重做日志文件的位置。默认为数据库默认的归档路径。

11.4　有归档日志的数据库不完全恢复

11.4.1　数据库不完全恢复概述

1．概念

有归档日志的前提下，不完全恢复是指数据库没有恢复到控制文件记录的最新时间点。这通常有两种情况导致的：一是归档日志文件有断裂时间点；二是用户特意要求恢复到过去某一时间点，如图 11-2 所示。

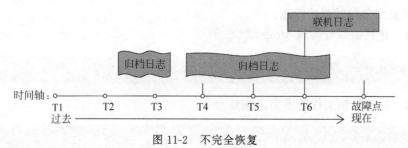

图 11-2　不完全恢复

2．数据库不完全恢复的注意事项

在进行数据库不完全恢复之前，首先要确保是否对数据库进行了完全备份。在进行数据文件损坏的不完全恢复时必须先使用完整的数据文件备份将数据库恢复到备份时刻的状态。

在不完全恢复后，需要使用 resetlogs 选项打开数据库，原来的重做日志被清空，新的重做日志文件序列号重新从 1 开始。

3. 数据库不完全恢复的分类

不完全恢复的类型可以归纳为如下所示。

(1) 基于时间的不完全恢复：将数据库恢复到备份与故障时刻之间的某个特定时刻。

(2) 基于撤销的不完全恢复：数据库的恢复随用户输入 cancel 命令而中止。

(3) 基于 SCN 的不完全恢复：将数据库恢复到指定的 SCN 值时的状态。

(4) 使用备份的控制文件进行恢复。

4. 数据库不完全恢复的语法

数据库不完全恢复的语法如下：

```
recover [automatic]
[from 'location'][database]
[until time time|cancel|change scn]
[using backup controlfile]
```

11.4.2 数据文件损坏的数据库不完全恢复

数据文件的损毁现象很常见，一旦遇到数据文件被损毁，就需及时采用如下操作步骤进行恢复。

步骤 1 如果数据库没有关闭，则强制关闭数据库。

```
SQL> shutdown abort
```

步骤 2 用备份的所有数据文件还原当前数据库的所有数据文件，即将数据库的所有数据文件恢复到备份时刻的状态。

步骤 3 将数据库启动到 Mount 状态。

```
SQL> startup mount
```

步骤 4 执行数据文件的不完全恢复命令。

```
SQL> recover database until time time：(基于时间恢复)。
SQL> recover database until cancel：(基于撤销恢复)。
SQL> recover database until change scn：(基于 SCN 恢复)。
```

步骤 5 可以通过查询数据字典视图 v$log_history 以获得时间和 SCN 的信息。

步骤 6 不完全恢复完成后，使用 resetlogs 选项启动数据库。

```
SQL> alter database open resetlogs;
```

11.4.3 控制文件损坏的数据库不完全恢复

步骤1 如果数据库没有关闭,则强制关闭数据库。

```
SQL> shutdown abort
```

步骤2 用备份的所有数据文件和控制文件还原当前数据库的所有数据文件、控制文件,即将数据库的所有数据文件、控制文件恢复到备份时刻的状态。

步骤3 将数据库启动到 Mount 状态。

```
SQL> startup mount
```

步骤4 执行不完全恢复命令。

```
SQL> recover database until time time using backup controlfile;
SQL> recover database until cancel using backup controlfile;
SQL> recover database until change scn using backup controlfile;
```

步骤5 不完全恢复完成后,使用 resetlogs 选项启动数据库。

```
SQL> alter database open resetlogs;
```

11.4.4 日志文件损坏的数据库不完全恢复

步骤1 如果数据库没有关闭,则强制关闭数据库。

```
SQL> shutdown abort
```

步骤2 将数据库启动到 Mount 状态。

```
SQL> startup mount
```

步骤3 执行日志文件的不完全恢复命令。

```
SQL> alter database clear log file group 日志组号;(非当前日志损坏的恢复)
SQL> recover database until cancel(当前日志损坏的非强制恢复)
SQL> alter system set "_allow_resetlogs_corruption" = true scope = spfile;(当前日志损坏的强制恢复)
```

步骤4 打开数据库并重置日志序号。

```
SQL> alter database open resetlogs;
```

11.5 项目案例

11.5.1 非归档模式下的备份与完全恢复实例

脱机全库数据库备份是一致性的备份,是很珍贵的数据库备份。下面演示非归档模式的备份与完全恢复的具体操作步骤与源码。

视频讲解

1. 模拟数据库损毁过程的操作步骤

首先在步骤1~步骤7中给出了实验的准备操作,即在表空间 USER 中创建实验用表 test2,然后依次插入数据,模拟 users01.dbf 文件被意外损毁。

步骤1 登录与打开数据库。

```
SQL> show user;
user 为 "SYS"
```

步骤2 创建实验用表,插入一行数据。

```
SQL> create table test2(a int);
表已创建.

SQL> insert into test2 values(1);
已创建 1 行.

SQL> commit;
提交完成.
```

步骤3 由于脱机备份需要在数据库关闭的状态下进行,所以应先关闭数据库。

```
SQL> shutdown immediate;
数据库已经关闭.
已经卸载数据库.
Oracle 例程已经关闭.
```

步骤4 执行冷备份操作。

```
SQL> @D:\BACKUP_NEWORCL\BACKUPNEWORCL\COOLBAK.SQL
已连接到空闲例程.
ERROR:
ORA - 01034: ORACLE not available
ORA - 27101: shared memory realm does not exist
进程 ID: 0
```

会话 ID: 0 序列号: 0

D:\Oracle19c\oradata\NEWORCL\CONTROL01.CTL
D:\Oracle19c\oradata\NEWORCL\CONTROL02.CTL
D:\Oracle19c\oradata\NEWORCL\REDO01.LOG
D:\Oracle19c\oradata\NEWORCL\REDO02-2.LOG
D:\Oracle19c\oradata\NEWORCL\REDO02.LOG
D:\Oracle19c\oradata\NEWORCL\REDO03-1.LOG
D:\Oracle19c\oradata\NEWORCL\REDO03-2.LOG
D:\Oracle19c\oradata\NEWORCL\REDO03-3.LOG
D:\Oracle19c\oradata\NEWORCL\REDO03.LOG
D:\Oracle19c\oradata\NEWORCL\STUDENT01.DBF
D:\Oracle19c\oradata\NEWORCL\SYSAUX01.DBF
D:\Oracle19c\oradata\NEWORCL\SYSTEM01.DBF
D:\Oracle19c\oradata\NEWORCL\TEMP01.DBF
D:\Oracle19c\oradata\NEWORCL\UNDOTBS01.DBF
D:\Oracle19c\oradata\NEWORCL\USERS01.DBF
复制了 15 个文件.

D:\BACKUP\CONTROL03.CTL
复制了 1 个文件.

步骤 5 启动数据库。

请输入用户名： sys as sysdb
输入口令：
已连接到空闲例程.

SQL> startup
Oracle 例程已经启动.

Total System Global Area 2550136752 bytes
Fixed Size   9031600 bytes
Variable Size   570425344 bytes
Database Buffers   1962934272 bytes
Redo Buffers   7745536 bytes
数据库装载完毕.
数据库已经打开.

步骤 6 插入第 2 行数据并提交,此时查询表 test 应该有两行数据。

QL> select * from test2;

         A

         1

```
SQL> insert into test2 values(2);
已创建 1 行.

SQL> commit;
提交完成.

SQL> select * from test2;

         A
----------
         1
         2
```

步骤 7　关闭数据库,删除文件 users01.dbf,以模拟数据文件丢失。

```
SQL> shutdown immediate
数据库已经关闭.
已经卸载数据库.
Oracle 例程已经关闭.
SQL> host del D:\Oracle19c\oradata\NEWORCL\users01.dbf
```

2. 全库数据库恢复的操作步骤

步骤 1　启动数据库,发现文件丢失,只能启动到装载阶段。

```
SQL> startup
ORACLE 例程已经启动.

Total System Global Area 2550136752 bytes
Fixed Size                    9031600 bytes
Variable Size               570425344 bytes
Database Buffers           1962934272 bytes
Redo Buffers                  7745536 bytes
数据库装载完毕.
ORA - 01157: 无法标识/锁定数据文件 7 - 请参阅 DBWR 跟踪文件
ORA - 01110: 数据文件 7: 'D:\ORACLE19C\ORADATA\NEWORCL\USERS01.DBF'
```

启动后,发现错误提示。这代表数据库已发现文件丢失,所以系统只能启动到装载 Mount 阶段,无法打开(Open)。

步骤 2　将之前的冷备份完全复制过来。由于进行的是全库恢复,控制文件也会恢复成旧的,所以需要将数据库关闭(当前数据库在装载阶段)。

```
SQL> shutdown immediate
ORA - 01109: 数据库未打开

已经卸载数据库.
Oracle 例程已经关闭.
```

此时提示数据库未打开,但是系统依然会关闭当前数据库。

步骤 3 在数据库关闭后,就可以将备份的内容进行恢复。

这个步骤可以使用命令,也可以手动将备份的数据文件、控制文件以及日志文件进行复制。

步骤 4 打开数据库到 Open 状态。

```
SQL> alter database open;

数据库已更改.
```

在启动数据库到 Open 状态这个阶段,有可能遇到很多小问题,比如闪回区大小问题等。下面给大家演示这个问题的解决办法。

```
SQL> startup
Oracle 例程已经启动.

Total System Global Area  2550136752 bytes
Fixed Size                   9031600 bytes
Variable Size              570425344 bytes
Database Buffers          1962934272 bytes
Redo Buffers                 7745536 bytes
数据库装载完毕.
ORA - 38760: 此数据库实例无法启用闪回数据库.

SQL> alter database flashback off;

数据库已更改.

SQL> alter database open;

数据库已更改.

SQL> alter database flashback on;

数据库已更改.
```

步骤 5 查看表 test2,观察数据情况。

```
SQL> select * from test2;

         A
----------
         1
```

在这个解决方案中需要用冷备份的全库文件进行恢复,所以能够恢复到冷备份时的一致性状态下。但是在表 test2 中只能观察到 1,而查询不到 2。

3. 只恢复单独数据文件的操作步骤

在上述案例中,大家可能想到,是否只复制损毁文件的备份即可,因为只有文件 users01.dbf 损毁。下面给出无归档日志的脱机备份与只恢复损坏文件 users01.dbf 的详细步骤:

步骤 1 启动数据库。

```
SQL> startup
Oracle 例程已经启动.

Total System Global Area 2550136752 bytes
Fixed Size                  9031600 bytes
Variable Size             570425344 bytes
Database Buffers         1962934272 bytes
Redo Buffers                7745536 bytes
数据库装载完毕.
ORA-01157: 无法标识/锁定数据文件 7 - 请参阅 DBWR 跟踪文件
ORA-01110: 数据文件 7: 'D:\ORACLE19C\ORADATA\NEWORCL\USERS01.DBF'
```

步骤 2 将冷备份的 users01.dbf 恢复到原位置后,恢复数据库。
这步可以采用命令进行文件复制,也可以手动复制,本节不再具体演示操作代码。

步骤 3 通过控制文件恢复数据库。
输入 auto 自动进行数据库恢复,但是数据库无法自动恢复。

```
SQL> recover database until cancel using backup controlfile;
ORA-00279: 更改 3409104 (在 02/18/2020 16:52:11 生成) 对于线程 1 是必需的
ORA-00289: 建议: D:\BACKUP\ARC0000000001_1032711158.0001
ORA-00280: 更改 3409104 (用于线程 1) 在序列 #1 中


指定日志: {<RET>= suggested | filename | AUTO | CANCEL}
auto
ORA-00308: 无法打开归档日志 'D:\BACKUP\ARC0000000001_1032711158.0001'
ORA-27041: 无法打开文件
OSD-04002: 无法打开文件
O/S-Error: (OS 2) 系统找不到指定的文件.
```

步骤 4 确认当前重做日志序号,以便利用指定日志文件进行恢复。

```
SQL> select group#,sequence#,members,status from v$log;

    GROUP#   SEQUENCE#    MEMBERS   STATUS
   --------- ----------- --------- -----------
        1         1          1     CURRENT
        3         0          3     UNUSED
        2         0          2     UNUSED
```

步骤 5 用 recover 手动输入恢复需要用的联机日志,再次恢复数据库,使用日志组 1 的日志文件。

```
SQL> recover database until cancel using backup controlfile;
ORA-00279: 更改 3409104 (在 02/18/2020 16:52:11 生成) 对于线程 1 是必需的
ORA-00289: 建议: D:\BACKUP\ARC0000000001_1032711158.0001
ORA-00280: 更改 3409104 (用于线程 1) 在序列 #1 中

指定日志: {<RET>= suggested | filename | AUTO | CANCEL}
D:\ORACLE19C\ORADATA\NEWORCL\redo01.log
已应用的日志.
完成介质恢复.
```

步骤 6 以重置日志文件的方式进行数据库重启。

```
SQL> alter database open resetlogs;

数据库已更改.
```

步骤 7 查看表空间 USERS 中表 test2 中的数据。

```
SQL> select * from test2;

         A
   ----------
         1
         2
```

11.5.2 联机备份下数据文件损毁的恢复实例

1. 联机备份下损坏一个数据文件的完全恢复的操作步骤

要进行数据库的联机备份,数据库必须运行在归档模式下。数据库在联机打开供用户

视频讲解

使用的情况下进行备份,使用的是以表空间为单位分段脱机备份的方式。每个表空间脱机备份后再联机时,该表空间中数据文件时间点就一定会落后了,因此,必须依赖归档日志对时间点进行调整。因为数据库的日志信息微观而详细地记录了数据库运行的全部过程。

步骤1　确定是否为归档模式,其代码如下:

```
SQL> archive log list;
数据库日志模式              存档模式
自动存档                    启用
存档终点                    d:\backup\
最早的联机日志序列           1
下一个存档日志序列           1
当前日志序列                 1
```

步骤2　创建实验用表 test3 并插入数据,其代码如下:

```
SQL> create table test3(a int) tablespace users;
表已创建.

SQL> insert into test3 values(1);
已创建 1 行.

SQL> commit;
提交完成.
```

步骤3　启动热备份,此时忽略所有热备份文件过程。可以直接运行热备份脚本实现热备份,其代码如下:

```
SQL> @D:\BACKUP_NEWORCL\BACKUPNEWORCL\HOTBAK.SQL
已连接.
系统已更改.
系统已更改.
表空间已更改.
D:\Oracle19c\oradata\NEWORCL\SYSTEM01.DBF
复制了 1 个文件.
表空间已更改.
表空间已更改.
D:\Oracle19c\oradata\NEWORCL\UNDOTBS01.DBF
复制了 1 个文件.
表空间已更改.
表空间已更改.
D:\Oracle19c\oradata\NEWORCL\SYSAUX01.DBF
复制了 1 个文件.
```

```
表空间已更改.
表空间已更改.
D:\Oracle19c\oradata\NEWORCL\USERS01.DBF
复制了 1 个文件.
表空间已更改.
表空间已更改.
D:\Oracle19c\oradata\NEWORCL\STUDENT01.DBF
复制了 1 个文件.
表空间已更改.
数据库已更改.
数据库已更改.
系统已更改.

    FILE#  STATUS           CHANGE#   TIME              CON_ID
    -----  ---------------  --------  ----------------  ------
        1  NOT ACTIVE        3417610  18-2月 -20             0
        2  NOT ACTIVE        3417692  18-2月 -20             0
        3  NOT ACTIVE        3417657  18-2月 -20             0
        4  NOT ACTIVE        3417640  18-2月 -20             0
        7  NOT ACTIVE        3417675  18-2月 -20             0
从 Oracle Database 19c Enterprise Edition Release 19.0.0.0.0 - Production
Version 19.3.0.0.0 断开
```

步骤 4　继续插入数据并提交,其代码如下:

```
SQL> insert into test3 values(2);
已创建 1 行.

SQL> commit;
提交完成.
```

步骤 5　强制进行联机日志归档,这样插入的第 2 行数据已经进入归档日志中,其代码如下:

```
SQL> alter system archive log current;

系统已更改.

SQL> alter system archive log current;

系统已更改.
```

步骤 6　关闭数据库,模拟文件丢失,其代码如下:

```
SQL> shutdown immediate;
数据库已经关闭.
已经卸载数据库.
```

```
Oracle 例程已经关闭.
SQL> host del D:\ORACLE19C\ORADATA\NEWORCL\USERS01.DBF;
```

步骤 7　若要进行数据库恢复,则先启动数据库。由于丢失文件,所以只能启动到装载阶段,其代码如下:

```
SQL> startup
Oracle 例程已经启动.

Total System Global Area 2550136752 bytes
Fixed Size                  9031600 bytes
Variable Size             570425344 bytes
Database Buffers         1962934272 bytes
Redo Buffers                7745536 bytes
数据库装载完毕.
ORA - 01157: 无法标识/锁定数据文件 7 - 请参阅 DBWR 跟踪文件
ORA - 01110: 数据文件 7: 'D:\ORACLE19C\ORADATA\NEWORCL\USERS01.DBF'
```

步骤 8　查询损坏的文件编号,其代码如下:

```
SQL> select * from v$recover_file;

    FILE#    ONLINE  ONLINE_ERROR             CHANGE#         TIME     CON_ID
--------- -------- ------------------------ --------------- -------- ---------
        7    ONLINE  ONLINE                   FILE NOT FOUND       0         0
```

步骤 9　将损坏的文件脱机,并打开数据库,其代码如下:

```
SQL> alter database datafile 7 offline drop;

数据库已更改.

SQL> alter database open;

数据库已更改.
```

步骤 10　复制备份的数据文件(restore)。

步骤 11　进行文件恢复。由于在前面进行了日志的切换和归档,所以在恢复的时候需要借助归档日志。输入 auto 命令可自动查找归档日志并进行恢复,其代码如下:

```
SQL> recover datafile 7;
ORA - 00279: 更改 3417675 (在 02/18/2020 19:02:21 生成) 对于线程 1 是必需的
ORA - 00289: 建议: D:\BACKUP\ARC0000000004_1032714364.0001
```

```
ORA-00280：更改 3417675（用于线程 1）在序列 #4 中

指定日志：{<RET>= suggested | filename | AUTO | CANCEL}
auto
已应用的日志.
完成介质恢复.
```

步骤 12 在恢复损坏的数据文件后，将数据文件 7 联机，并查询表 test3。

```
SQL> alter database datafile 7 online;

数据库已更改.

SQL> select * from test3;

        A
----------
        1
        2
```

如果在恢复的过程中，发生损坏的是多个数据文件，可以采用逐一对数据文件进行恢复的方法，如上述的步骤 9 需要对数据文件一一脱机，步骤 10 需要对数据文件进行一一恢复；也可以采用整个数据库的恢复方法。接下来的案例展示的就是整个数据库的恢复方法。

2. 全库损毁后数据库的不完全恢复的操作步骤

在归档模式下损坏或丢失多个数据文件，可进行整个数据库的恢复。

步骤 1 登录并创建实验用表 test4，其代码如下：

```
SQL> create table test4(a int) tablespace users;
表已创建.

SQL> insert into test4 values(1);
已创建 1 行.

SQL> commit;
提交完成.
```

步骤 2 联机数据库，进行数据文件的备份（热备份），其代码如下：

```
SQL> @D:\BACKUP_NEWORCL\BACKUPNEWORCL\HOTBAK.SQL
已连接.
系统已更改.
系统已更改.
表空间已更改.
```

```
D:\Oracle19c\oradata\NEWORCL\SYSTEM01.DBF
复制了 1 个文件
表空间已更改.
表空间已更改.
D:\Oracle19c\oradata\NEWORCL\UNDOTBS01.DBF
复制了 1 个文件
表空间已更改.
表空间已更改.
D:\Oracle19c\oradata\NEWORCL\SYSAUX01.DBF
复制了 1 个文件
表空间已更改.
表空间已更改.
D:\Oracle19c\oradata\NEWORCL\USERS01.DBF
复制了 1 个文件
表空间已更改.
表空间已更改.
D:\Oracle19c\oradata\NEWORCL\STUDENT01.DBF
复制了 1 个文件
表空间已更改.
数据库已更改.
数据库已更改.
系统已更改.

    FILE#  STATUS          CHANGE#  TIME            CON_ID
    -----  --------------  -------  --------------  ------
        1  NOT ACTIVE      3417610  18-2月 -20           0
        2  NOT ACTIVE      3417692  18-2月 -20           0
        3  NOT ACTIVE      3417657  18-2月 -20           0
        4  NOT ACTIVE      3417640  18-2月 -20           0
        7  NOT ACTIVE      3417675  18-2月 -20           0
从 Oracle Database 19c Enterprise Edition Release 19.0.0.0.0 - Production
Version 19.3.0.0.0 断开
```

步骤 3 继续插入一行实验数据并提交,将当前日志归档,其代码如下:

```
SQL> insert into test4 values(2);

已创建 1 行.

SQL> commit;

提交完成.

SQL> alter system archive log current;

系统已更改.
```

步骤 4 再插入一行数据,提交,自动归档,其代码如下:

```
SQL> insert into test4 values(3);
```

已创建 1 行.

```
SQL> commit;
```

提交完成.

步骤 5　关闭当前数据库,其代码如下:

```
SQL> shutdown
```
数据库已经关闭.
已经卸载数据库.
Oracle 例程已经关闭.

步骤 6　模拟数据库损坏,删除全库的 3 类文件,其代码如下:

```
SQL> host del D:\ORACLE19C\ORADATA\NEWORCL\ *.*;
D:\ORACLE19C\ORADATA\NEWORCL\ *.*,是否确认(Y/N)? y
```

步骤 7　启动数据库。由于没有控制文件,所以只能启动到第一阶段,其代码如下:

```
SQL> startup
Oracle 例程已经启动.

Total System Global Area 2550136752 bytes
Fixed Size                    9031600 bytes
Variable Size               570425344 bytes
Database Buffers           1962934272 bytes
Redo Buffers                  7745536 bytes
ORA - 00205: ?????????, ??????, ???????
```

以下为进行不完全恢复的操作步骤。

步骤 8　查看初始化参数文件中的控制文件名称以及路径,其代码如下:

```
SQL> show parameter control;

NAME                                 TYPE               VALUE
------------------------------------ ------------------ ---------------
control_file_record_keep_time        integer            7
control_files                        string
D:\ORACLE19C\ORADATA\NEWORCL\CONTROL01.CTL,              D:\ORACLE19C\ORA
DATA\NEWORCL\CONTROL02.CTL, D:\BACKUP\CONTROL03.CTL
control_management_pack_access       string             DIAGNOSTIC + TUNING
```

步骤9　按照以上文件中的路径和名称,依次复制备份的控制文件,其代码如下:

```
SQL> host xcopy D:\BACKUP\CONTROL0202.BAK
D:\ORACLE19C\ORADATA\NEWORCL\CONTROL01.CTL
目标 D:\ORACLE19C\ORADATA\NEWORCL\CONTROL01.CTL 是文件名还是目录名
(F = 文件,D = 目录)?F
D:\Backup\CONTROL0202.BAK 复制了一个文件
```

步骤10　数据库打开到装载阶段,其代码如下:

```
SQL> alter database mount;

数据库已更改.
```

查询 v$ercover_file 视图可以得到待恢复的文件名,也可以查询告警文件以观察丢失的文件,其代码如下:

```
SQL> select * from v$recover_file;

FILE#      ONLINE   ONLINE_ERROR       CHANGE#  TIME                    CON_ID
------     ------   -------------      -------  ---------------------   ------
     1     ONLINE   ONLINE   FILE NOT FOUND          0                       0
     2     ONLINE   ONLINE   FILE NOT FOUND          0                       0
     3     ONLINE   ONLINE   FILE NOT FOUND          0                       0
     4     ONLINE   ONLINE   FILE NOT FOUND          0                       0
     7     ONLINE   ONLINE   FILE NOT FOUND          0                       0
```

步骤11　将以前备份的数据文件复制过来。可以手动进行复制,也可以使用命令进行恢复,其代码如下:

```
SQL> host xcopy    备份文件    原地点
```

步骤12　查询联机日志的状态,其代码如下:

```
SQL> select group#,sequence#,status,first_change# from v$log;

   GROUP#   SEQUENCE#  STATUS       FIRST_CHANGE#
   ------   ---------  ----------   -------------
         1          7  INACTIVE           3417869
         3          6  INACTIVE           3417865
         2          8  CURRENT            3420487
```

此时的联机日志状态是从原控制文件中读取的,记录的当前的连接日志序号为8。而在热备份数据以后,步骤3又强行归档了一次日志。因此数据库在故障时的当前序号应为9,而9号联机日志发生损坏,所以只能恢复到8号日志的末尾处。以上查询的最后一个字段,即 first_change# 就是联机日志最早的 SCN 号。

步骤 13　执行不完全恢复。系统需根据各文件的 SCN 自动定位需要重做的日志。在此实例中,系统首先选定需要重做的 8 号日志,选择系统建议的日志重做即可,其代码如下:

```
SQL> recover database using backup controlfile until cancel;
ORA - 00279: ?? 3420629 (? 02/18/2020 19:47:03 ??) ???? 1 ????
ORA - 00289: ??: D:\BACKUP\ARC0000000010_1032714364.0001
ORA - 00280: ?? 3420629 (???? 1) ??? #10 ?

指定日志: {<RET>= suggested | filename | AUTO | CANCEL}
cancel
介质恢复已取消.
```

步骤 14　打开数据库。因为数据库的联机日志不好用了,必须重置联机日志序号,系统才能从控制文件中得到日志文件等信息,自动生成 3 组联机日志文件,其代码如下:

```
SQL> alter database open resetlogs;

数据库已更改.
```

此时查询联机日志信息,其代码如下:

```
SQL> select group#,sequence#,status,first_change# from v$log;

    GROUP#   SEQUENCE#   STATUS           FIRST_CHANGE#
    ------   ---------   --------------   -------------
         1           1   CURRENT                3420630
         2           0   UNUSED                       0
         3           0   UNUSED                       0
```

步骤 15　查询表 test4,发现数据 1、2 存在,而数据 3 丢失,其代码如下:

```
SQL> select * from test4;

         A
  --------
         1
         2
```

3. 只损毁了数据文件而进行的恢复的操作步骤

在全库损毁后数据库的不完全恢复的操作步骤 6 中,如果只删除了多个数据文件,而保留了控制文件和日志文件,那么恢复的操作就简单得多了,而且可以实现完全恢复。下面简单阐述如下所述。

步骤1　启动数据库,检查错误。既可以查看告警文件,也可以使用动态性能视图以确定丢失文件。

步骤2　复制备份数据文件回到原地点,开始恢复数据库,其代码语法如下:

SQL> host xcopy　备份文件　原地点
SQL> recover database;在恢复过程中选择 auto 即可。

步骤3　打开数据库,检查数据库的数据是否完全恢复。

11.5.3　损毁日志文件的恢复实例

1. 损毁非当前工作日志的恢复操作步骤

步骤1　查看当前日志状态,发现当前联机日志组为2,其代码如下:

```
SQL> select group#, sequence#, bytes, blocksize, members, archived, status
  2  from v$log;

    GROUP#  SEQUENCE#      BYTES  BLOCKSIZE    MEMBERS ARC STATUS
---------- ---------- ---------- ---------- ---------- --- ----------
         1         10  209715200        512          1 YES INACTIVE
         2         12  209715200        512          2 NO  CURRENT
         3         11   15728640        512          3 YES ACTIVE
```

步骤2　关闭数据库,模拟日志文件损坏,删除非当前联机日志,例如第一组日志,其代码如下:

```
SQL> shutdown
数据库已经关闭.
已经卸载数据库.
Oracle 例程已经关闭.
SQL> host del D:\ORACLE19C\ORADATA\NEWORCL\REDO01.LOG;
```

步骤3　打开数据库,发现数据库不能正常启动,其代码如下:

```
SQL> startup
Oracle 例程已经启动.

Total System Global Area 2550136752 bytes
Fixed Size                   9031600 bytes
Variable Size              570425344 bytes
Database Buffers          1962934272 bytes
Redo Buffers                 7745536 bytes
```

数据库装载完毕.
ORA－03113：通信通道的文件结尾
进程 ID: 72552
会话 ID: 373 序列号: 17363

此时数据库运行在加载状态下。当然上面的启动数据库语句也可以直接定义为：
SQL>startup mount
在发现问题后，可以查看告警日志观察具体的问题所在，如图 11-2 所示。

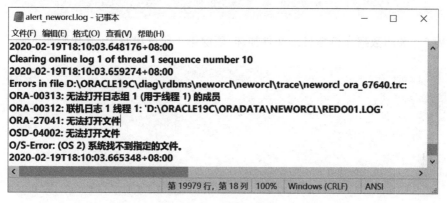

图 11-2　告警文件示意

步骤 4　清除日志文件组 1，其代码如下：

SQL>alter database clear logfile group 1;

数据库已更改.

如果该日志组没有归档，则需要使用如下代码：

SQL>alter database clear unarchived logfile group 1;

数据库已更改.

步骤 5　更改数据库状态为打开，其代码如下：

SQL>alter database open;

数据库已更改.

步骤 6　重新备份数据库。

2. 损毁当前日志的强制性恢复的操作步骤

如果当前损毁的日志为当前活动日志，最好的办法就是通过不完全恢复，可以保证数据库的一致性，但是这种办法要求在归档方式下，并且有可用的备份。当然通过强制性恢复也可以，但是可能会导致数据库的不一致。

步骤1　查看日志,发现当前日志为日志组1,其代码如下:

```
SQL> select group#,sequence#,bytes,blocksize,members,archived,status from v$log;

    GROUP#  SEQUENCE#       BYTES  BLOCKSIZE    MEMBERS ARC STATUS
    ------  ---------  ----------  ---------  --------- --- --------
         1         16   209715200        512          1 NO  CURRENT
         2         15   209715200        512          2 YES ACTIVE
         3         14    15728640        512          3 YES INACTIVE
```

步骤2　关闭数据库模拟日志文件损坏,删除第1组联机日志,其代码如下:

```
SQL> shutdown immediate
数据库已经关闭.
已经卸载数据库.
Oracle 例程已经关闭.

SQL> host del   D:\ORACLE19C\ORADATA\NEWORCL\REDO01.LOG
```

步骤3　重新打开数据库,发现缺失日志文件,无法打开,其代码如下:

```
SQL> startup
Oracle 例程已经启动.

Total System Global Area 2550136752 bytes
Fixed Size                  9031600 bytes
Variable Size             570425344 bytes
Database Buffers         1962934272 bytes
Redo Buffers                7745536 bytes
数据库装载完毕.
ORA-03113: 通信通道的文件结尾
进程 ID: 53356
会话 ID: 250 序列号: 19674
```

步骤4　进行强制性恢复。首先输入 recover database until cancel 命令告诉数据库恢复的终点。先选择 auto,尽量恢复可以利用的归档日志,然后重新输入 cancel,完成不完全恢复,也就是说要进行两次恢复的操作。

```
SQL> recover database until cancel;

完成介质恢复.
```

步骤5　打开数据库并重置日志序号,其代码如下:

```
SQL> alter database open resetlogs;
```

数据库已更改.

注意:
(1) 这种办法恢复的数据库是一致的不完全恢复,会丢失当前联机日志中的事务数据。
(2) 这种方法适合于归档数据库并且有可用的数据库全备份。
(3) 恢复成功之后,记得再做一次数据库的全备份。
(4) 建议联机日志文件一定要实现镜像在不同的磁盘上,避免这种情况的发生,因为任何数据的丢失对于生产来说都是不容许的。

11.5.4 损坏控制文件的恢复实例

视频讲解

控制文件是 Oracle 数据库中一类非常重要的文件,在进行备份和恢复的操作时,它的重要性尤为突出。在进行数据库的恢复操作时,一般以控制文件的 SCN 为恢复的目标,所以控制文件损坏意味着这个目标的丢失,会带来很多问题。

在数据库启动阶段,了解到数据库要想启动到装载(Mount)阶段,必须读取控制文件。因此,如果控制文件损坏,数据库就只能停留在启动的第一阶段——非装载(Nomount)阶段。

数据库中有多个控制文件,这些控制文件之间是镜像的关系,所以,对于单个控制文件损坏的恢复是很简单的,只需用未损坏的控制文件替换损坏的控制文件即可;对于控制文件被全部损坏的情况,备份的控制文件的时间点一般比当前数据库晚,不宜使用,可以采用重新创建控制文件的方式来进行恢复。

1. 损坏单个控制文件的恢复的操作步骤

首先模拟单个控制文件的损毁,然后启动数据库观察问题,具体的操作步骤如下所示。

步骤 1 模拟单个控制文件被损坏,最典型的就是启动数据库时显示出错,且不能启动数据库到 Mount 状态。

```
SQL> startup
ORA-00205: error in identifying controlfile, check alert log for more info
查看告警日志文件,有如下信息
alter database   mount
Mon May 26 11:59:52 2003
ORA-00202: controlfile: 'D:\Oracle19c\oradata\NEWORCL\control01.ctl'
ORA-27041: unable to open file
OSD-04002: unable to open file
O/S-Error: (OS 2) 系统找不到指定的文件.
```

如果想要查看具体的错误提示,就可以检查告警日志,如图 11-3 所示。

图 11-3 告警日志示意

步骤 2　停止当前数据库。

步骤 3　复制一个好的控制文件以替换坏的控制文件,或修改 init.ora 中的控制文件参数以取消这个坏的控制文件,如 alter system set control_files = '[value]' scope = both;

步骤 4　重新启动数据。

注意:

(1) 损失单个控制文件是比较简单的,因为数据库中所有的控制文件都是镜像的,只需要简单的复制一个就可以了。

(2) 建议将镜像控制文件保存在不同的磁盘上。

(3) 建议多做几个控制文件的备份,并长期保留一份由 alter database backup controlfile to trace 产生的控制文件的文本备份。

2. 全部控制文件被损坏的恢复的操作步骤

步骤 1　进行数据库控制文件的备份,其代码如下:

```
SQL> alter database backup controlfile to trace;
数据库已更改.


SQL> select tracefile from v$process where addr in (select paddr from v$session where sid in
 (select sid from v$mystat));


TRACEFILE
--------------------------------------------------------------------------------
D:\ORACLE19C\diag\rdbms\neworcl\neworcl\trace\neworcl_ora_74252.trc
```

步骤 2　关闭数据库,模拟丢失了全部控制文件,其代码如下:

```
SQL> shutdown immediate;
数据库已经关闭.
已经卸载数据库.
Oracle 例程已经关闭.
```

步骤 3　启动数据库,显示错误,只能启动到非装载阶段(Nomount)。具体错误可以查看 alert log 文件,其代码如下:

```
SQL> startup
Oracle 例程已经启动.

Total System Global Area 2550136752 bytes
Fixed Size                  9031600 bytes
Variable Size             570425344 bytes
Database Buffers         1962934272 bytes
Redo Buffers                7745536 bytes
ORA-00205: ?????????, ??????, ???????
```

观察告警日志文件可发现具体的错误信息,如图 11-4 所示。

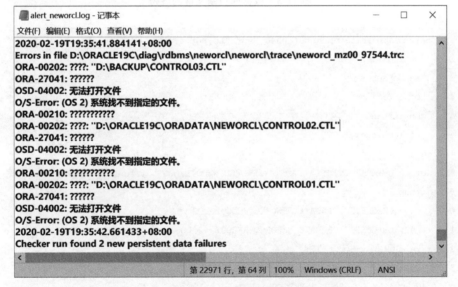

图 11-4　告警日志示意

步骤 4　关闭数据库,运行步骤 1 所备份的控制文件脚本。

(1) 根据 v$process 与 v$mystat 动态性能视图,可以查看到数据库控制文件的备份,具体如图 11-5 所示。

(2) 复制并执行该脚本,创建控制文件,其代码如下:

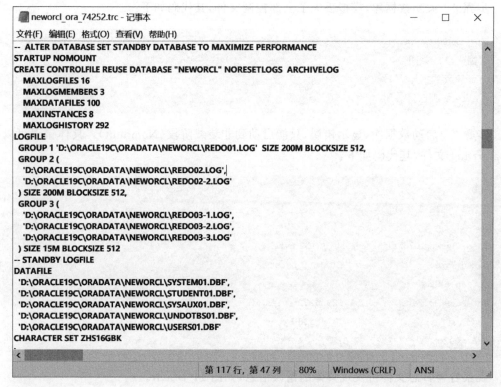

图 11-5 数据库控制文件的备份

```
SQL> create controlfile reuse database "NEWORCL" noresetlogs archivelog
  2     maxlogfiles   16
  3     maxlogmebers   3
  4     maxdatafiles   100
  5     maxinstances   8
  6     maxloghistory   292
  7  logfile
  8     group 1 'D:\ORACLE19C\ORADATA\NEWORCL\REDO01.LOG'   size 200M blocksize 512,
  9     group 2 (
 10        'D:\ORACLE19C\ORADATA\NEWORCL\REDO02.LOG',
 11        'D:\ORACLE19C\ORADATA\NEWORCL\REDO02 - 2.LOG'
 12     )size 200M blocksize 512,
 13     group 3 (
 14        'D:\ORACLE19C\ORADATA\NEWORCL\REDO03 - 1.LOG',
 15        'D:\ORACLE19C\ORADATA\NEWORCL\REDO03 - 2.LOG',
 16        'D:\ORACLE19C\ORADATA\NEWORCL\REDO03 - 3.LOG'
 17     ) size 15M blocksize 512
 18     —standby logfile
 19     datafile
```

```
20      'D:\ORACLE19C\ORADATA\NEWORCL\SYSTEM01.DBF',
21      'D:\ORACLE19C\ORADATA\NEWORCL\STUDENT01.DBF',
22      'D:\ORACLE19C\ORADATA\NEWORCL\SYSAUX01.DBF',
23      'D:\ORACLE19C\ORADATA\NEWORCL\UNDOTBS01.DBF',
24      'D:\ORACLE19C\ORADATA\NEWORCL\USERS01.DBF'
25   character set ZHS16GBK;

控制文件已创建.
```

步骤 5　输入 until cancel 的 recover 命令进行恢复。

由于数据库除了控制文件被损坏外,其他文件全部完好,故不需要恢复。在这里使用 until cancel 命令只是为了告诉大家,数据库已经到了恢复的终点,所以输入这个命令后,直接输出 cancel 即可。

```
SQL> recover database using backup controlfile until cancel;
ORA-00279: ?? 3394378 (? 02/19/2020 19:32:43 ??) ???? 1 ????
ORA-00289: ??: D:\BACKUP\ARC0000000001_1032806747.0001
ORA-00280: ?? 3394378 (???? 1) ??? #1 ?

指定日志: {<RET>= suggested | filename | AUTO | CANCEL}
cancel
介质恢复已取消.
```

步骤 6　修改数据库至打开状态。

最后将数据库打开,由于控制文件是刚创建的文件,其中的日志序号时间点信息与别的文件不一致,所以为了达到一致,需要使用 restlogs 来重新设置日志序号为 1,其代码如下:

```
SQL> alter database open resetlogs;

数据库已更改.
```

注意:

(1) 重建控制文件用于恢复全部数据文件的损坏,需要注意其书写的正确性,保证包含了所有的数据文件与联机日志。

(2) 经常有这样一种情况:因为一个磁盘被损坏了,就不能再恢复数据文件到这个磁盘,因此在恢复到另外一个磁盘的时候就必须重新创建控制文件,以用于识别这个新的数据文件。可以用上述方法用于恢复数据。

11.5.5　基于时间点的不完全恢复实例

在实际的应用中,经常有需要将某个数据文件恢复到过去某个时间点的要求。例如,用户误删了某个重要的表,需要恢复到这个用户操作之前的时间点。特别注意的是:恢复到过去某个时间点需要以在这个时间点之前的某个实体备份为基础,然后重做这个备份到待恢复时

视频讲解

间点之间的归档日志；而绝对不是从正在运行的数据库直接回到过去的某个时间点。

基于时间点的不完全恢复的操作步骤如下所示。

步骤1　登录与启动数据库。

步骤2　创建实验表 test5 并插入实验数据提交。

```
SQL> create table test5(a int);
表已创建.

SQL> insert into test5 values(1);
已创建 1 行.

SQL> commit;
提交完成.
```

步骤3　热备份数据库。其可以直接调用热备份脚本文件，其代码如下：

```
SQL> @ D:\BACKUP_NEWORCL\BACKUPNEWORCL\HOTBAK.SQL
已连接.
系统已更改.
系统已更改.
表空间已更改.
D:\Oracle19c\oradata\NEWORCL\SYSTEM01.DBF
复制了 1 个文件
表空间已更改.
表空间已更改.
D:\Oracle19c\oradata\NEWORCL\UNDOTBS01.DBF
复制了 1 个文件
表空间已更改.
表空间已更改.
D:\Oracle19c\oradata\NEWORCL\SYSAUX01.DBF
复制了 1 个文件
表空间已更改.
表空间已更改.
D:\Oracle19c\oradata\NEWORCL\USERS01.DBF
复制了 1 个文件
表空间已更改.
表空间已更改.
D:\Oracle19c\oradata\NEWORCL\STUDENT01.DBF
复制了 1 个文件
表空间已更改.
数据库已更改.
数据库已更改.
系统已更改.
```

```
    FILE#  STATUS              CHANGE#    TIME           CON_ID
    -----  ------------------  ---------  ------------   --------
        1  NOT ACTIVE          3410456    20-2月 -20      0
        2  NOT ACTIVE          3410530    20-2月 -20      0
        3  NOT ACTIVE          3410495    20-2月 -20      0
        4  NOT ACTIVE          3410478    20-2月 -20      0
        7  NOT ACTIVE          3410513    20-2月 -20      0
```

步骤 4　再插入一行新的数据,并对当前日志归档,其代码如下:

```
SQL> insert into test5 values(2);
已创建 1 行.

SQL> commit;
提交完成.

SQL> alter system archive log current;
系统已更改.
```

步骤 5　插入一行新数据,并提交,无需手动归档,其代码如下:

```
SQL> insert into test5 values(3);
已创建 1 行.

SQL> commit;
提交完成.
```

步骤 6　查询当前系统时间,其代码如下:

```
SQL> select to_char(sysdate,'yyyy-mm-dd hh24:mi:ss') from dual;

TO_CHAR(SYSDATE,'YY
-------------------
2020-02-20 17:53:37
```

步骤 7　模拟表 test5 的误操作,其代码如下:

```
SQL> drop table test5;
表已删除.
```

以下的操作步骤开始进行数据库的恢复。

步骤 8　关闭数据库并将所有数据文件复制过来,其代码如下:

```
SQL> shutdown
数据库已经关闭.
已经卸载数据库.
Oracle 例程已经关闭.
```

依次将热备份的数据文件复制至 oradata 文件夹即可,这里就不再演示数据文件复制的过程了。

步骤 9 数据库启动到装载(Mount)阶段,并开始恢复到特定的时间点,其代码如下:

```
SQL> startup mount;
Oracle 例程已经启动.
Total System Global Area 2550136752 bytes
Fixed Size                   9031600 bytes
Variable Size              570425344 bytes
Database Buffers          1962934272 bytes
Redo Buffers                 7745536 bytes
数据库装载完毕.

SQL> recover database until time '2020 - 02 - 20 17:53:37';
完成介质恢复.
```

步骤 10 打开数据库,查询表 test5,显示 3 个数据已全部恢复,其代码如下:

```
SQL> alter database open resetlogs;

数据库已更改.

SQL> select * from test5;
         A
----------
         1
         2
         3
```

注意:

(1) 在进行不完全恢复操作时最好备份所有的数据,冷备份也可以。因为恢复过程是从备份点往后恢复的。如果因为其中一个数据的时间戳(SCN)大于要恢复的时间点,那么恢复都是不可能成功的。

(2) 在不完全恢复之后,都必须使用 resetlogs 方式打开数据库,并马上再做一次全备份,因为 resetlogs 之后再用以前的备份恢复就很难了。

(3) 以上是在删除之前获得时间,但在实际应用中,很难知道删除之前的实际时间,但可以采用大致时间,或可以采用分析日志文件取得精确的需要恢复的时间。

(4)一般都是在测试机或备用机器上采用这种不完全恢复,在恢复之后通过导出/导入被误删的表以回到生产系统。

11.5.6 基于 SCN 的不完全恢复实例

在 11.5.5 节中给出基于时间的不完全恢复,除了这种方式以外,也可以基于系统改变号 SCN 将数据库恢复到过去的某个 SCN。这只需要在 11.5.5 节中更改两个操作步骤即可。

(1)在步骤 6 中,将查询语句修改为查询 SCN,其代码如下:

```
SQL> select dbms_flashback.get_system_change_number from dual;

GET_SYSTEM_CHANGE_NUMBER
------------------------
                 3412202
```

(2)在步骤 9 中,语句更改为"recover database until change SCN",其代码如下:

```
SQL> recover database until change  3412202;

已应用的日志.
完成介质恢复.
```

第12章 Oracle数据的移动

PPT 视频讲解

12.1 逻辑备份与逻辑恢复

逻辑备份是指利用 Oracle 提供的导出工具,将数据库中选定的记录集或数据字典的逻辑副本以二进制文件的形式存储到操作系统中。逻辑恢复是指利用 Oracle 提供的导入工具将逻辑备份形成的转储文件导入数据库内部,进行数据库的逻辑恢复。逻辑备份与逻辑恢复必须在数据库运行的状态下进行。因此,数据库备份与恢复是以物理备份与物理恢复为主,以逻辑备份与逻辑恢复为辅。

EXP 和 IMP 是 Oracle 比较经典也是比较古老的逻辑备份和逻辑恢复的方式。而数据泵技术即 Data Pump Export(EXPDP)和 Data Pump Import(IMPDP)实用程序,它可以实现数据的导出与导入。

注意:这两类逻辑备份与逻辑恢复实用程序之间不兼容。使用 EXP 备份的转储文件,不能使用 IMPDP 进行导入;同样,使用 EXPDP 备份的转储文件,也不能使用 IMP 进行导入。

12.2 数据泵技术

1. 数据泵技术的概念

数据泵技术相对应的工具是 Data Pump Export 和 Data Pump Import。它的功能与前面介绍的 EXP 和 IMP 类似;所不同的是,数据泵高速并行的设计使得服务器在运行导入和导出任务时更加快速,可以装载或卸载大量数据。另外,数据泵可以实现断点重启,即一个任务无论是人为地中断还是意外中断,都可以从断点地方重新启动。所以数据泵技术主要用于对大量数据的大作业操作。在使用数据泵进行数据导出与导入时,可以使用多线程并行操作。

2. 数据泵的目录对象

数据泵在数据库服务器上可创建所有的备份文件，而 Oracle 系统要求数据泵必须使用目录对象，以防止用户误操作数据库服务器上特定目录下的操作系统文件。目录对象对应于操作系统上的一个指定目录。

（1）查询默认的目录对象的工作目录，命令如下所示。

```
SQL> select * from dba_directories where directory_name = 'DATA_PUMP_DIR';

OWNER     DIRECTORY_NAME      DIRECTORY_PATH                        ORIGIN_CON_ID
--------- ------------------- ------------------------------------- --------------
SYS       DATA_PUMP_DIR       D:\Oracle19c\admin\neworcl\dpdump\    0
```

（2）创建目录对象权限。如果用户想要使用数据泵命令，但是没有可用的目录，也不具备创建目录的权限，就会提示相应的错误。所以如果用户需要自己创建目录对象，就必须具有创建任何目录（create any directory）的权限。当然也可以先在用户 sys 或 system 下创建数据泵导入/导出目录，然后将该目录赋予其他用户使用，即将该目录的读和写权限赋予用户即可，其代码如下：

```
SQL> create or replace directory dumpdir as 'D:\BACKUP';

目录已创建.

SQL> grant read, write on directory dumpdir to hr;

授权成功.
```

如果用户要导出或导入非同名模式的对象，还需要具有 EXP_FULL_DATABASE 和 IMP_FULL_DATABASE 权限，其代码如下：

```
SQL> grant exp_full_database, imp_full_database to hr;

授权成功.
```

12.3 数据泵导出

12.3.1 EXPDP 导出概述

EXPDP 将数据库中的元数据与行数据导出到操作系统的转储文件中。

1. EXPDP 导出工具的执行方式

（1）命令行方式：在命令行中直接指定参数设置。

(2) 参数文件方式：将参数设置存放到一个参数文件中，在命令行中用 parfile 参数指定参数文件。

(3) 交互方式：通过交互式命令进行导出作业管理。

2. 数据泵导出模式

(1) 全库导出：通过参数 full 指定，导出整个数据库。

(2) 模式导出：通过参数 schemas 指定，是默认的导出模式，导出指定模式中的所有对象。

(3) 表导出：通过参数 tables 指定，导出指定模式中指定的所有表、分区及其依赖对象。

(4) 表空间导出：通过参数 tablespaces 指定，导出指定表空间中所有表及其依赖对象的元数据和行数据。

(5) 传输表空间导出：通过参数 transport_tablespaces 指定，导出指定表空间中所有表及其依赖对象的元数据。

12.3.2 EXPDP 参数介绍

EXPDP 的具体参数，可查询帮助文档，执行效果如下所示。

```
C:\WINDOWS\system32>expdp help=y

Export: Release 19.0.0.0.0 - Production on 星期五 2 月 21 12:28:27 2020
Version 19.3.0.0.0

Copyright (c) 1982, 2019, Oracle and/or its affiliates.  All rights reserved.
数据泵导出实用程序提供了一种用于在 Oracle 数据库之间传输
数据对象的机制.该实用程序可以使用命令进行调用:

   示例: expdp scott/tiger DIRECTORY=dmpdir DUMPFILE=scott.dmp
您可以控制导出的运行方式.具体方法是: 在 'expdp' 命令后输入
各种参数.要指定各参数, 请使用关键字:

   格式:  expdp KEYWORD=value 或 KEYWORD=(value1,value2,...,valueN)
   示例: expdp scott/tiger DUMPFILE=scott.dmp DIRECTORY=dmpdir SCHEMAS=scott
            或 TABLES=(T1:P1,T1:P2), 如果 T1 是分区表
```

常用的参数介绍如下所述。

(1) content：指定要导出的内容。

(2) data_only：表示只导出对象的行数据。

(3) metadata_only：表示只导出对象的元数据。

(4) directory：指定转储文件和日志文件所在位置的目录对象，该对象由 DBA 预先创建。

(5) dumpfile：指定转储文件名称列表，可以包含目录对象名。默认值为 expdat.dmp。

(6) full：指定是否进行全数据库导出，包括所有行数据与元数据。

(7) job_name：指定导出作业的名称。默认值为系统自动为作业生成的一个名称。

(8) logfile：指定导出日志文件的名称。默认值为 export.log。

(9) parallel：指定执行导出作业时最大并行进程个数。默认值为 1。
(10) parfile：指定参数文件的名称。
(11) schemas：指定进行模式导出及模式名称列表。
(12) tables：指定进行表模式导出及表名称列表。
(13) tablespaces：指定进行表空间模式导出及表空间名称列表。
(14) transport_tablespaces：指定进行传输表空间模式导出及表空间名称列表。

12.3.3　EXPDP 导出实例

下面以用户 system 登录，分别使用下面 4 个简单的导出语句进行表、模式、表空间以及全库 4 种模式导出。

1. 表导出

```
D:\> expdp system/neworcl@neworcl dumpfile = data_pump_dir:hr_tables_%u.dat tables = hr.employees,hr.departments nologfile = y

Export: Release 19.0.0.0.0 - Production on 星期五 2月 21 18:11:10 2020
Version 19.3.0.0.0

Copyright (c) 1982, 2019, Oracle and/or its affiliates.  All rights reserved.

连接到: Oracle Database 19c Enterprise Edition Release 19.0.0.0.0 - Production
启    动 "SYSTEM"."SYS_EXPORT_TABLE_01":  system/********@neworcl dumpfile = DATA_PUMP_DIR:hr_tables_%u.dat tables = hr.employees,hr.departments nologfile = y
处理对象类型 TABLE_EXPORT/TABLE/TABLE_DATA
处理对象类型 TABLE_EXPORT/TABLE/INDEX/STATISTICS/INDEX_STATISTICS
处理对象类型 TABLE_EXPORT/TABLE/STATISTICS/TABLE_STATISTICS
处理对象类型 TABLE_EXPORT/TABLE/STATISTICS/MARKER
处理对象类型 TABLE_EXPORT/TABLE/TABLE
处理对象类型 TABLE_EXPORT/TABLE/COMMENT
处理对象类型 TABLE_EXPORT/TABLE/INDEX/INDEX
处理对象类型 TABLE_EXPORT/TABLE/CONSTRAINT/CONSTRAINT
处理对象类型 TABLE_EXPORT/TABLE/CONSTRAINT/REF_CONSTRAINT
处理对象类型 TABLE_EXPORT/TABLE/TRIGGER
. . 导出了 "HR"."EMPLOYEES"              17.08 KB    107 行
. . 导出了 "HR"."DEPARTMENTS"             7.125 KB    27 行
已成功加载/卸载了主表 "SYSTEM"."SYS_EXPORT_TABLE_01"
******************************************************************************
SYSTEM.SYS_EXPORT_TABLE_01 的转储文件集为:
  D:\ORACLE19C\ADMIN\NEWORCL\DPDUMP\HR_TABLES_01.DAT
作业 "SYSTEM"."SYS_EXPORT_TABLE_01" 已于 星期五 2月 21 18:11:48 2020 elapsed 0 00:00:36 成功完成
```

2. 模式导出

```
D:\>expdp  system/neworcl@neworcl  dumpfile = data_pump_dir:hr_schema.dmp  logfile = DATA_PUMP_DIR:hr_schema.log

Export: Release 19.0.0.0.0 - Production on 星期五 2月 21 18:02:43 2020
Version 19.3.0.0.0

Copyright (c) 1982, 2019, Oracle and/or its affiliates.  All rights reserved.

连接到: Oracle Database 19c Enterprise Edition Release 19.0.0.0.0 - Production
启动 "SYSTEM"."SYS_EXPORT_SCHEMA_01":  system/******** @neworcl
dumpfile = DATA_PUMP_DIR:hr_schema.dmp logfile = DATA_PUMP_DIR:hr_schema.log
处理对象类型 SCHEMA_EXPORT/TABLE/TABLE_DATA
处理对象类型 SCHEMA_EXPORT/TABLE/INDEX/STATISTICS/INDEX_STATISTICS
处理对象类型 SCHEMA_EXPORT/TABLE/STATISTICS/TABLE_STATISTICS
处理对象类型 SCHEMA_EXPORT/DEFAULT_ROLE
处理对象类型 SCHEMA_EXPORT/PRE_SCHEMA/PROCACT_SCHEMA
处理对象类型 SCHEMA_EXPORT/TABLE/TABLE
处理对象类型 SCHEMA_EXPORT/TABLE/COMMENT
处理对象类型 SCHEMA_EXPORT/TABLE/INDEX/INDEX
已成功加载/卸载了主表 "SYSTEM"."SYS_EXPORT_SCHEMA_01"
******************************************************************************
SYSTEM.SYS_EXPORT_SCHEMA_01 的转储文件集为:
   D:\ORACLE19C\ADMIN\NEWORCL\DPDUMP\HR_SCHEMA.DMP
作业 "SYSTEM"."SYS_EXPORT_SCHEMA_01" 已于 星期五 2月 21 18:03:51 2020 elapsed 0 00:01:05 成功完成
```

3. 表空间导出

```
D:\>expdp system/neworcl@neworcl dumpfile = data_pump_dir:users_tbs_%u.dmp  tablespaces = users parallel = 2 logfile = DATA_PUMP_DIR:user_tbs.log

Export: Release 19.0.0.0.0 - Production on 星期五 2月 21 18:08:11 2020
Version 19.3.0.0.0

Copyright (c) 1982, 2019, Oracle and/or its affiliates.  All rights reserved.

连接到: Oracle Database 19c Enterprise Edition Release 19.0.0.0.0 - Production
启动 "SYSTEM"."SYS_EXPORT_TABLESPACE_01":  system/******** @neworcl
dumpfile = DATA_PUMP_DIR:users_tbs_%u.dmp tablespaces = users parallel = 2
logfile = DATA_PUMP_DIR:user_tbs.log
```

```
处理对象类型 TABLE_EXPORT/TABLE/STATISTICS/TABLE_STATISTICS
处理对象类型 TABLE_EXPORT/TABLE/TABLE_DATA
处理对象类型 TABLE_EXPORT/TABLE/TABLE
处理对象类型 TABLE_EXPORT/TABLE/AUDIT_OBJ
处理对象类型 TABLE_EXPORT/TABLE/STATISTICS/MARKER
. . 导出了 "U1"."TEST"                                          0 KB        0 行
已成功加载/卸载了主表 "SYSTEM"."SYS_EXPORT_TABLESPACE_01"
******************************************************************************
SYSTEM.SYS_EXPORT_TABLESPACE_01 的转储文件集为：
  D:\ORACLE19C\ADMIN\NEWORCL\DPDUMP\USERS_TBS_01.DMP
  D:\ORACLE19C\ADMIN\NEWORCL\DPDUMP\USERS_TBS_02.DMP
作业 "SYSTEM"."SYS_EXPORT_TABLESPACE_01" 已于 星期五 2 月 21 18:08:30 2020 elapsed 0 00:00:
16 成功完成
```

4. 全库导出

```
D:\>expdp system/neworcl@neworcl dumpfile = data_pump_dir:mydb_%u.dat nologfile = y full = y

Export: Release 19.0.0.0.0 - Production on 星期五 2 月 21 17:42:31 2020
Version 19.3.0.0.0

Copyright (c) 1982, 2019, Oracle and/or its affiliates.  All rights reserved.

连接到: Oracle Database 19c Enterprise Edition Release 19.0.0.0.0 - Production
启动 "SYSTEM"."SYS_EXPORT_FULL_01":   system/********@neworcl
dumpfile = DATA_PUMP_DIR:mydb_%u.dat nologfile = y full = y
处理对象类型 DATABASE_EXPORT/EARLY_OPTIONS/VIEWS_AS_TABLES/TABLE_DATA
处理对象类型 DATABASE_EXPORT/NORMAL_OPTIONS/TABLE_DATA
. . . . . . .
. . 导出了 "SYS"."KU$_USER_MAPPING_VIEW"               6.078 KB      38 行
. . 导出了 "AUDSYS"."AUD$UNIFIED":"SYS_P181"           87.60 KB      80 行
. . 导出了 "AUDSYS"."AUD$UNIFIED":"SYS_P887"           76.81 KB      76 行
. . 导出了 "AUDSYS"."AUD$UNIFIED":"SYS_P550"           52.74 KB       4 行
. . . . . . .
已成功加载/卸载了主表 "SYSTEM"."SYS_EXPORT_FULL_01"
******************************************************************************
SYSTEM.SYS_EXPORT_FULL_01 的转储文件集为：
  D:\ORACLE19C\ADMIN\NEWORCL\DPDUMP\MYDB_01.DAT
作业 "SYSTEM"."SYS_EXPORT_FULL_01" 已于 星期五 2 月 21 17:48:06 2020 elapsed 0 00:05:33 成功
完成
```

12.4 数据泵导入

12.4.1 IMPDP 导入概述

数据泵导入是一个用于将转储文件导入目标数据库的工具。它可以将转储文件导入到源数据库中，也可以导入到其他平台上运行的不同版本的 Oracle 数据库中。

数据泵导入的执行也可以采用交互方式、命名行方式以及参数文件方式 3 种。

数据泵导入的导入模式也可以归纳为以下 4 种。

(1) 全库导入：将源数据库的所有元数据与行数据都导入到目标数据库。

(2) 模式导入：通过参数 schema 指定，将指定模式中所有对象的元数据与行数据导入目标数据库。

(3) 表导入：通过参数 tables 指定，将指定表、分区以及依赖对象导入目标数据库。

(4) 表空间导入：通过参数 tablespaces 指定，将指定表空间中所有对象及其依赖对象的元数据和行数据导入目标数据库。

12.4.2 IMPDP 参数介绍

使用 impdp-help 命令可查看数据泵导入参数，执行效果如下所示。

```
C:\WINDOWS\system32>impdp help = y

Import: Release 19.0.0.0.0 - Production on 星期五 2 月 21 12:31:16 2020
Version 19.3.0.0.0

Copyright (c) 1982, 2019, Oracle and/or its affiliates.  All rights reserved.

数据泵导入实用程序提供了一种用于在 Oracle 数据库之间传输
数据对象的机制.该实用程序可以使用以下命令进行调用:
示例: impdp scott/tiger DIRECTORY = dmpdir DUMPFILE = scott.dmp

您可以控制导入的运行方式.具体方法是: 在 'impdp' 命令后输入
各种参数.要指定各参数,请使用关键字:
    格式:   impdp KEYWORD = value 或 KEYWORD = (value1,value2,...,valueN)
    示例: impdp scott/tiger DIRECTORY = dmpdir DUMPFILE = scott.dmp
```

常用的参数介绍如下所述。

(1) content：指定要导入的内容。

(2) data_only：只导入对象的行数据。

(3) metadata_only：只导入对象的元数据。

（4）directory：指定转储文件和日志文件所在位置的目录对象，该对象由 DBA 预先创建。

（5）dumpfile：指定转储文件名称列表，可以包含目录对象名，默认值为 expdat.dmp。

（6）full：指定是否进行全数据库导入，包括所有元数据与行数据。

（7）include：指定导入操作中要导入的对象类型和对象元数据。

（8）job_name：指定导入作业的名称。默认值为系统自动为作业生成的一个名称。

（9）logfile：指定导入日志文件的名称。

（10）nologfile：是否生成导入日志。

（11）parallel：执行导入作业时并行进程的最大个数。

（12）parfile：指定参数文件的名称。

（13）query：指定导入操作中 SELECT 语句中的数据导入条件。

（14）remap_schema：将源模式中的所有对象导入到目标模式中。

（15）remap_table：允许在导入操作过程中重命名表。

（16）remap_tablespace：将源表空间所有对象导入目标表空间中。

（17）schemas：指定进行模式导入的模式名称列表。默认为当前用户模式。

（18）tables：指定表模式导入的表名称列表。

（19）tablespaces：指定进行表空间模式导入的表空间名称列表。

（20）transport_tablespaces：指定进行传输表空间模式导入的表空间名称列表。

12.4.3　IMPDP 导入实例

下面以用户 system 登录，分别使用下面 4 个简单的导入语句进行表、模式、表空间、以及全库 4 种模式导入。

（1）表导入。

```
D:\>impdp system/neworcl@neworcl dumpfile = data_pump_dir:hr_tables_01.dat nologfile = y
tables = hr.employees

Import: Release 19.0.0.0.0 - Production on 星期五 2 月 21 18:34:11 2020
Version 19.3.0.0.0

Copyright (c) 1982, 2019, Oracle and/or its affiliates.  All rights reserved.

连接到: Oracle Database 19c Enterprise Edition Release 19.0.0.0.0 - Production
已成功加载/卸载了主表 "SYSTEM"."SYS_IMPORT_TABLE_01"
启动 "SYSTEM"."SYS_IMPORT_TABLE_01":  system/******** @neworcl
dumpfile = DATA_PUMP_DIR:HR_TABLES_01.dat nologfile = y tables = hr.employees
处理对象类型 TABLE_EXPORT/TABLE/TABLE
```

```
ORA-39151: 表 "HR"."EMPLOYEES" 已存在. 由于跳过了 table_exists_action, 将跳过所有相关元数
据和数据.

处理对象类型 TABLE_EXPORT/TABLE/TABLE_DATA
处理对象类型 TABLE_EXPORT/TABLE/COMMENT
处理对象类型 TABLE_EXPORT/TABLE/INDEX/INDEX
处理对象类型 TABLE_EXPORT/TABLE/CONSTRAINT/CONSTRAINT
处理对象类型 TABLE_EXPORT/TABLE/INDEX/STATISTICS/INDEX_STATISTICS
处理对象类型 TABLE_EXPORT/TABLE/CONSTRAINT/REF_CONSTRAINT
处理对象类型 TABLE_EXPORT/TABLE/TRIGGER
处理对象类型 TABLE_EXPORT/TABLE/STATISTICS/TABLE_STATISTICS
处理对象类型 TABLE_EXPORT/TABLE/STATISTICS/MARKER
作业 "SYSTEM"."SYS_IMPORT_TABLE_01" 已经完成, 但是有 1 个错误 (于 星期五 2 月 21 18:34:19
2020 elapsed 0 00:00:05 完成)
```

(2) 模式导入。

```
D:\>impdp system/neworcl@neworcl dumpfile=data_pump_dir:hr_schema.dmp  nologfile=y
include=table,trigger


Import: Release 19.0.0.0.0 - Production on 星期五 2 月 21 18:36:12 2020
Version 19.3.0.0.0

Copyright (c) 1982, 2019, Oracle and/or its affiliates.   All rights reserved.


连接到: Oracle Database 19c Enterprise Edition Release 19.0.0.0.0 - Production
ORA-31655: 尚未为作业选择数据或元数据对象
已成功加载/卸载了主表 "SYSTEM"."SYS_IMPORT_FULL_01"
启动 "SYSTEM"."SYS_IMPORT_FULL_01":   system/********@neworcl
dumpfile=DATA_PUMP_DIR:HR_SCHEMA.dmp nologfile=y include=table,trigger
作业 "SYSTEM"."SYS_IMPORT_FULL_01" 已于 星期五 2 月 21 18:36:16 2020 elapsed 0 00:00:03 成功完成
```

(3) 表空间导入。

```
D:\>impdp system/neworcl@neworcl dumpfile=data_pump_dir:users_tbs_01.dmp,users_tbs_02.
dmp tablespaces=users


Import: Release 19.0.0.0.0 - Production on 星期五 2 月 21 18:30:05 2020
Version 19.3.0.0.0

Copyright (c) 1982, 2019, Oracle and/or its affiliates.   All rights reserved.


连接到: Oracle Database 19c Enterprise Edition Release 19.0.0.0.0 - Production
已成功加载/卸载了主表 "SYSTEM"."SYS_IMPORT_TABLESPACE_01"
```

启动 "SYSTEM"."SYS_IMPORT_TABLESPACE_01": system/********@neworcl
dumpfile=DATA_PUMP_DIR:USERS_TBS_01.dmp,USERS_TBS_02.dmp tablespaces=users
处理对象类型 TABLE_EXPORT/TABLE/TABLE
ORA-39151: 表 "U1"."TEST" 已存在。由于跳过了 table_exists_action，将跳过所有相关元数据和数据。

处理对象类型 TABLE_EXPORT/TABLE/TABLE_DATA
处理对象类型 TABLE_EXPORT/TABLE/AUDIT_OBJ
处理对象类型 TABLE_EXPORT/TABLE/STATISTICS/TABLE_STATISTICS
处理对象类型 TABLE_EXPORT/TABLE/STATISTICS/MARKER
作业 "SYSTEM"."SYS_IMPORT_TABLESPACE_01" 已经完成，但是有 1 个错误（于 星期五 2月 21 18:30:13 2020 elapsed 0 00:00:06 完成）

（4）数据库导入。

```
D:\>impdp system/neworcl@neworcl dumpfile=data_pump_dir:mydb_01.dat parallel=3 full=y

Import: Release 19.0.0.0.0 - Production on 星期五 2月 21 18:23:21 2020
Version 19.3.0.0.0

Copyright (c) 1982, 2019, Oracle and/or its affiliates.  All rights reserved.

连接到: Oracle Database 19c Enterprise Edition Release 19.0.0.0.0 - Production
已成功加载/卸载了主表 "SYSTEM"."SYS_IMPORT_FULL_01"
启动 "SYSTEM"."SYS_IMPORT_FULL_01":  system/********@neworcl
dumpfile=DATA_PUMP_DIR:MYDB_01.dat parallel=3 full=y
处理对象类型 DATABASE_EXPORT/PRE_SYSTEM_IMPCALLOUT/MARKER
处理对象类型 DATABASE_EXPORT/PRE_INSTANCE_IMPCALLOUT/MARKER
处理对象类型 DATABASE_EXPORT/TABLESPACE
ORA-31684: 对象类型 TABLESPACE:"UNDOTBS1" 已存在
……
作业 "SYSTEM"."SYS_IMPORT_FULL_01" 已经完成，但是有 31 个错误（于 星期五 2月 21 18:24:44 2020 elapsed 0 00:01:21 完成）
```

12.5 项目案例

12.5.1 逻辑备份与逻辑恢复实例

视频讲解

步骤 1　数据库转存目录。在使用数据泵导出和数据泵导入程序之前需要创建 directory 目录对象。

（1）利用已有的默认导入导出工作目录，其代码如下：

```
SQL> select * from dba_directories;

OWNER       DIRECTORY_NAME              DIRECTORY_PATH                              ORIGIN_CON_ID
----------------------------------------------------------------------------------------------
SYS         DUMPDIR                     D:\BACKUP                                   0
SYS         ORACLECLRDIR                D:\software\oracle19c\bin\clr               0
SYS         SDO_DIR_WORK                                                            0
SYS         SDO_DIR_ADMIN               D:\software\oracle19c/md/admin              0
SYS         XMLDIR                      D:\software\oracle19c\rdbms\xml             0
SYS         XSDDIR                      D:\software\oracle19c\rdbms\xml\schema      0
SYS         OPATCH_INST_DIR             D:\software\oracle19c\OPatch                0
SYS         ORACLE_OCM_CONFIG_DIR2      D:\software\oracle19c\ccr\state             0
SYS         ORACLE_BASE                 D:\Oracle19c                                0
SYS         ORACLE_HOME                 D:\software\oracle19c                       0
SYS         ORACLE_OCM_CONFIG_DIR       D:\software\oracle19c\ccr\state             0
SYS         DATA_PUMP_DIR               D:\Oracle19c\admin\neworcl\dpdump/          0
SYS         OPATCH_SCRIPT_DIR           D:\software\oracle19c\QOpatch               0
SYS         OPATCH_LOG_DIR              D:\software\oracle19c\rdbms\log             0
SYS         JAVA$JOX$CUJS$DIRECTORY$    D:\SOFTWARE\ORACLE19C\JAVAVM\ADMIN\         0

已选择 15 行.
```

（2）data_pump_dir 存储的为系统默认目录，可以使用默认目录，也可以自行创建数据库目录，例如本章中创建的 dumpdir。具体创建目录 dumpdir 的代码如下：

```
SQL> create or replace directory dumpdir as 'D:\BACKUP';

目录已创建.
```

步骤 2　为用户授予权限。

（1）将该目录对象的读和写权限授予用户。

（2）如果用户导出完全数据库，就需要用户具有以下权限：exp_full_database 和 imp_full_database 权限。

```
SQL> grant exp_full_database,imp_full_database to hr;

授权成功.


SQL> grant read,write on directory dumpdir to hr;

授权成功.
```

步骤 3　在用户 hr 创建测试用表，其代码如下：

```
SQL>create table emp_dump as select * from employees;
表已创建。

SQL>create table dept_dump as select * from departments;
表已创建。
```

下面演示利用数据泵执行数据导出的操作步骤。

步骤4 创建数据泵导出脚本文件。

为了使用方便,可以创建一个名为 hr.txt 的正文参数文件,如图 12-1 所示。

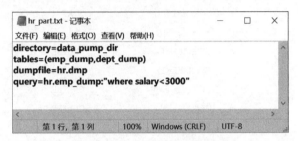

图 12-1 hr.txt 的正文参数文件

步骤5 执行该数据泵导出脚本,其代码如下:

```
D:\>expdp hr/hr parfile=hr.txt

Export: Release 19.0.0.0.0 - Production on 星期五 2 月 21 16:13:16 2020
Version 19.3.0.0.0
Copyright (c) 1982, 2019, Oracle and/or its affiliates.  All rights reserved.
连接到: Oracle Database 19c Enterprise Edition Release 19.0.0.0.0 - Production
启动 "HR"."SYS_EXPORT_TABLE_01":  hr/******** parfile=hr.txt
处理对象类型 TABLE_EXPORT/TABLE/TABLE_DATA
处理对象类型 TABLE_EXPORT/TABLE/STATISTICS/TABLE_STATISTICS
处理对象类型 TABLE_EXPORT/TABLE/STATISTICS/MARKER
处理对象类型 TABLE_EXPORT/TABLE/TABLE
. . 导出了 "HR"."EMP_DUMP"                            11.32 KB      26 行
. . 导出了 "HR"."DEPT_DUMP"                           7.125 KB      27 行
已成功加载/卸载了主表 "HR"."SYS_EXPORT_TABLE_01"
******************************************************************************
HR.SYS_EXPORT_TABLE_01 的转储文件集为:
D:\ORACLE19C\ADMIN\NEWORCL\DPDUMP\HR.DMP
作业 "HR"."SYS_EXPORT_TABLE_01" 于 星期五 2 月 21 16:13:51 2020 elapsed 0 00:00:34 成功完成
```

在执行过程中可能发现如下错误,如图 12-2 所示。

解决该问题的办法如下所述。

(1) 查询一下表空间的状态。既然上面的错误提示为临时表空间为空,那么首先查询

```
连接到: Oracle Database 19c Enterprise Edition Release 19.0.0.0.0 - Production
ORA-25153: 临时表空间为空
ORA-06512: 在 "SYS.DBMS_LOB", line 741
ORA-06512: 在 "SYS.DBMS_DATAPUMP", line 5420
ORA-06512: 在 line 1
```

图 12-2 错误提示

一下临时表空间,其代码如下:

```
SQL> select tablespace_name,contents,status from dba_tablespaces;

TABLESPACE_NAME                CONTENTS            STATUS
------------------------------ ------------------- ---------
SYSTEM                         PERMANENT           ONLINE
SYSAUX                         PERMANENT           ONLINE
UNDOTBS1                       UNDO                ONLINE
TEMP                           TEMPORARY           ONLINE
USERS                          PERMANENT           ONLINE
TEMP1                          TEMPORARY           ONLINE
STUDENT                        PERMANENT           ONLINE

已选择 7 行.
```

(2) 继续查询临时表空间的文件,其代码如下:

```
SQL> select tablespace_name,file_name from dba_temp_files;

未选定行
```

查询结果显示临时表空间确实为空。

(3) 设置临时表空间。手动设置临时表空间,增加临时表空间的数据文件 temp.dbf,其代码如下:

```
SQL> alter tablespace temp add tempfile '/ORADATA/ORCLGPS/TEMP.DBF' size   100m autoextend on next 10m;

表空间已更改
```

在解决完临时表空间的问题后,再次执行数据泵导出操作,操作可正常执行。

步骤 6 模拟错误,删除 emp_dump 表,其代码如下:

```
SQL> drop table hr.emp_dump;

表已删除.
```

第12章　Oracle数据的移动

步骤7　利用数据泵执行数据导入,其代码如下:

```
D:\>impdp hr/hr parfile=hr.txt

Import: Release 19.0.0.0.0 - Production on 星期五 2 月 21 17:03:50 2020
Version 19.3.0.0.0
Copyright (c) 1982, 2019, Oracle and/or its affiliates.  All rights reserved.
连接到: Oracle Database 19c Enterprise Edition Release 19.0.0.0.0 - Production
已成功加载/卸载了主表 "HR"."SYS_IMPORT_TABLE_01"
启动 "HR"."SYS_IMPORT_TABLE_01":  hr/******** parfile=hr.txt
处理对象类型 TABLE_EXPORT/TABLE/TABLE
ORA-39151: 表 "HR"."DEPT_DUMP" 已存在。由于跳过了 table_exists_action,将跳过所有相关元数据和数据。
处理对象类型 TABLE_EXPORT/TABLE/TABLE_DATA
. . 导入了 "HR"."EMP_DUMP"                        11.32 KB      26 行
处理对象类型 TABLE_EXPORT/TABLE/STATISTICS/TABLE_STATISTICS
处理对象类型 TABLE_EXPORT/TABLE/STATISTICS/MARKER
作业 "HR"."SYS_IMPORT_TABLE_01" 已经完成,但是有 1 个错误 (于 星期五 2 月 21 17:04:28 2020 elapsed 0 00:00:36 完成)
```

步骤8　验证 demp_dump 表是否恢复,其代码如下:

```
SQL> desc hr.emp_dump;
名称                                      是否为空?      类型
----------------------------------------- -------- ----------------
EMPLOYEE_ID                                          NUMBER(6)
FIRST_NAMEVARCHAR2                                   (20)
LAST_NAME                                 NOT NULL   VARCHAR2(25)
EMAIL                                     NOT NULL   VARCHAR2(25)
PHONE_NUMBER                                         VARCHAR2(20)
HIRE_DATE                                 NOT NULL   DATE
JOB_ID                                    NOT NULL   VARCHAR2(10)
SALARY                                               NUMBER(8,2)
COMMISSION_PCT                                       NUMBER(2,2)
MANAGER_ID                                           NUMBER(6)
DEPARTMENT_ID                                        NUMBER(4)
```

12.5.2　数据移动实例

若想将数据导出至 Excel 表中,则具体执行如下所示的操作步骤。

步骤1　在 D 盘根目录下创建 SQL 脚本文件 emp.sql,如图 12-3 所示。

视频讲解

图 12-3 emp.sql 文件

步骤 2 登录用户 hr，并运行脚本文件。

```
SQL> conn hr
输入口令:
已连接.
SQL>@D:\EMP.SQL
```

步骤 3 查看生成的正文文件，如图 12-4 所示。

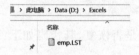

图 12-4 生成的正文文件

步骤 4 运行 Microsoft Excel，打开 emp.list 文件即可看到如图 12-5 所示结果。

	A	B	C	D	E	F	G	H	I	J	K	L
1												
2	EMPLOYEE_ID	FIRST_NAME	LAST_NAME	EMAIL		PHONE_NUMBER		HIRE_DATE		JOB_ID		SALARY
3												
4	COMMISSION_PCT	MANAGER_ID	DEPARTMENT_ID									
5												
6		100	Steven	King	SKING		515.123.4567		17-6月 -03	AD_PRES		23900
7					########							
8												
9		101	Neena	Kochhar	NKOCHHAR		515.123.4568		21-9月 -05	AD_VP		16900
10			100		########							
11												
12		102	Lex	De Haan	LDEHAAN		515.123.4569		13-1月 -01	AD_VP		16900
13			100		########							
14												
15		103	Alexander	Hunold	AHUNOLD		590.423.4567		03-1月 -06	IT_PROG		8900
16			102		########							
17												
18		104	Bruce	Ernst	BERNST		590.423.4568		21-5月 -07	IT_PROG		5900
19			103		########							
20												
21		105	David	Austin	DAUSTIN		590.423.4569		25-6月 -05	IT_PROG		4700
22			103		########							

图 12-5 emp.list 文件

第13章 Oracle闪回技术

PPT 视频讲解

为了使 Oracle 数据库从任何逻辑错误操作中迅速恢复,Oracle 数据库提供了一系列人为错误更正技术,称之为闪回(Flashback)。用闪回技术可以实现基于磁盘上闪回恢复的自动备份与恢复。

13.1 闪回技术

Oracle 数据库的闪回技术是一种数据恢复技术,具有恢复时间快,不使用备份文件的特点,能使数据库可以回到过去的某个状态,可以满足用户逻辑错误的快速恢复。

注意:闪回技术仅仅对逻辑恢复有效,如果是数据文件损坏,则必须使用介质恢复。

在没有闪回技术之前,数据库的逻辑错误恢复都是采用基于时间点,通过备份恢复数据库到过去指定的时间点的恢复,这种恢复方式需要使用备份并使用适当的归档日志完成,恢复时间取决于备份文件的复制时间和日志的应用时间。如果需要恢复的是很大的数据库,则可能恢复很长时间。若采用闪回技术,则可以更快速、更便捷地恢复用户错误或数据库逻辑错误。

13.2 闪回设置

1. 闪回恢复区的含义

Oracle 推荐指定一个闪回恢复区(Flashback Recovery Area)作为存放备份与恢复相关的默认位置,这样 Oracle 就可以实现自动的基于磁盘的备份与恢复。

闪回恢复区主要通过以下 3 个初始化参数来设置和管理。

(1) db_recovery_file_dest:指定闪回恢复区的位置。

(2) db_recovery_file_dest_size:指定闪回恢复区的可用空间。

(3) db_flashback_retention_target:该参数用来控制闪回日志中数据保留的时间,或者说,希望闪回数据库能够恢复到的最早的时间点。其单位为 min(分钟),默认是 1440min,即一天。

2. 闪回开启

开启闪回的代码如下所述。

```
SQL> shutdown immediate
数据库已经关闭.
已经卸载数据库.
Oracle 例程已经关闭.
SQL> startup mount
Oracle 例程已经启动.

Total System Global Area 2550136752 bytes
Fixed Size                   9031600 bytes
Variable Size             1191182336 bytes
Database Buffers          1342177280 bytes
Redo Buffers                 7745536 bytes
数据库装载完毕.
SQL> alter database archivelog;
数据库已更改.

SQL> alter database flashback on;
数据库已更改.

SQL> alter database open;
数据库已更改.
```

3. 确认闪回区设置

闪回开启后,可通过下列语句进行闪回区设置的查询。

```
SQL> show parameter db_recovery_file_dest;

NAME                                 TYPE        VALUE
------------------------------------ ----------- ------------------------------
db_recovery_file_dest                string      D:\Oracle19c\fast_recovery_area\
db_recovery_file_dest_size           big integer 500M
SQL> show parameter db_recovery_file_dest_size

NAME                                 TYPE        VALUE
------------------------------------ ----------- ------------------------------
db_recovery_file_dest_size           big integer 500M
SQL> show parameter db_flashback_retention_target
```

```
NAME                              TYPE           VALUE
------------------------------------------------------------
db_flashback_retention_target     integer        1440
```

4．闪回技术分类

具体的闪回技术包括以下 5 项。

（1）闪回查询：(Flashback Query)：查询过去某个时间点或某个 SCN 值时表中的数据信息。

（2）闪回版本查询(Flashback Version Query)：查询过去某个时间段或某个 SCN 段内表中数据变化的情况。

（3）闪回表(Flashback Table)：将表恢复到过去的某个时间点或某个 SCN 值时的状态。

（4）闪回删除(Flashback Drop)：将已经删除的表及其关联的对象恢复到删除前的状态。

（5）闪回数据库(Flashback Database)：将数据库恢复到过去某个时间点或某个 SCN 值时的状态。

13.3　闪回查询

闪回查询是查询该表过去某个时刻的数据情况，依赖于表空间 UNDO 的数据，一旦确认某个时刻的数据满足需求以后，可以根据这个时间执行闪回表，闪回查询先查询，等确认后再闪回到需求时刻。

闪回查询的基本语法如下：

```
select column_name[,...] from table_name
[as of SCN|timestamp expression]
[where conditoion]
```

13.4　闪回版本查询

所谓版本指的是每次事务所引起的数据行的变化情况，每一次变化就是一个版本。Oracle 提供了闪回版本查询，以此查询就可以看到数据行的整个变化过程。

闪回版本查询的基本语法如下：

```
select column_name[,... ] from table_name
versions between SCN|timestamp
minvalue|expresssion and maxvalue|expression
[as of SCN|timestamp expression]
where   condition
```

注意：闪回版本查询使用的是表空间 UNDO 里记录的 undo 数据，若 undo 段的数据由于空间压力而被清除，则产生无法闪回的情况。

13.5 闪回表

闪回表就是对表的数据做回退操作，取消对表所进行的修改，从而回退到之前的某个时间点。闪回表要求用户具有以下权限。

（1）flashback any table 权限或者是该表的 Flashback 对象权限。
（2）有该表的 select、insert、delete、alter 权限。
（3）必须保证该表 row movement。

要实现闪回表，就必须确保与撤销表空间有关的参数设置得合理。在 SQL Plus 中执行下面的语句以显示撤销表空间的参数。

```
SQL> show parameter undo

NAME                                 TYPE        VALUE
------------------------------------ ----------- ----------
temp_undo_enabled                    boolean     FALSE
undo_management                      string      AUTO
undo_retention                       integer     900
undo_tablespace                      string      UNDOTBS1
```

要对表进行闪回操作，就需要启动表的 row movement 特性。启动语句如下所示：

```
SQL> alter table table enable row movement;
```

闪回表操作的基本语法为

```
flashback table [schema.]table to SCN|timestamp
expression [enable|disable triggers]
```

其中：
schema：方案名称。
before drop：表示恢复到删除之前。
rename to table：表示恢复时更换表名。
SCN：SCN 是系统改变号，可以从 flashback_transaction_query 数据字典中查到。
timestamp：表示时间点，包含年月日以及时分秒。
enable disable triggers：表示触发器恢复之后的状态为 ENABLE。默认为 DISABLE 状态。

13.6 闪回删除

当用户对表进行 DDL(Date Definition Language，数据库模式定义语言)操作时，它是自动执行的。如果误删除了某个表，闪回就会为数据库提供一个安全机制——闪回删除。闪回删除可以恢复使用 drop table 语句删除的表。它是一种针对意外删除情况下的表恢复机制。

1. 回收站

闪回删除功能的实现主要是通过 Oracle 数据库中的"回收站"技术实现的。在 Oracle 数据库中,当执行 drop table 操作时,并不立即收回表及其关联对象的空间,而是将它们重命名后放入一个名为"回收站"的逻辑容器中保存,直到用户决定永久删除它们或存储该表的表空间或存储空间不足时,表才会被真正地删除。

回收站是一个虚拟容器,用于存储所有被删除的对象。为了避免被删除的表与同类对象名称重复,被删除的表(或者其他对象)放到回收站时,Oracle 系统对被删除表(或对象名)进行了转换。转换后的名称格式如下:

BIN $ globalUID $ Sversion

其中,globalUID 是一个全局唯一的标识对象,长度为 24 个字符,它是 Oracle 内部使用的标识。$ Sversion 是数据库分配的版本号。

为了使用数据库的闪回删除技术,必须开启数据库的"回收站"。当执行 drop table 操作时,表及其关联对象被命名后保存在"回收站"中,可以通过 user_recyclebin、dba_recyclebin 视图获得被删除的表及其关联对象信息。如果多个回收站记录具有相同原名称,则:使用系统生成的唯一名称来闪回特定版本。如果当前正在使用原名称,闪回的表遵循后进先出(LIFO)的规则重命名原名称。

回收站的工作原理如图 13-1 所示。

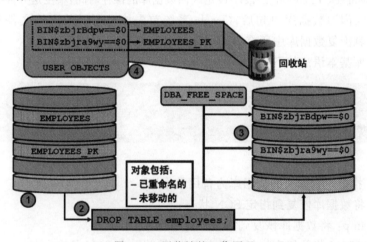

图 13-1 回收站的工作原理

但是如果在删除表时使用了 purge 短语,那么表以及关联对象则被直接释放,空间被回收,相关信息不会进入回收站。

使用 purge 短语的删除语句如下所示。

```
SQL> drop table<table_name>[purge];
```

2. 闪回删除

闪回删除的语法如下所示。

```
flashback table<table_name>
to before drop [rename to<new_name>];
```

由于被删除表及其关联对象的信息保存在回收站,其存储空间并没有释放,因此需要定期清空回收站,或清除回收站中没有用的对象(表、索引、表空间),以释放其所占磁盘空间。清除回收站的语法如下所示。

```
purge [table table|index index]
|[recyclebin|dba_recyclebin]
|[tablespace tablespace[user user]]
```

13.7 闪回数据库

闪回数据库是将数据库快速恢复到过去的某个时间点或 SCN 时的状态。当误删除一个用户或者误截断一个表时,可以采用数据库级的闪回恢复。

为了能在发生误操作时,闪回数据库到误操作之前的时间点上,需要设置下面 3 个参数。

(1) db_recovery_file_dest:确定 Flashback Logs 的存放路径。

(2) db_recovery_file_dest_size:指定恢复区的大小,默认值为空。

(3) db_flashback_retention_target:设定闪回数据库的保存时间,单位是分钟,默认是一天。

当用户发布闪回数据库语句后,Oracle 系统首先检查所需的归档文件和联机重做日志,如果正常,就恢复数据库中所有数据文件到指定的 SCN 或时间点上。

闪回数据库基本语法如下所述。

```
flashback [standby]database[database] to
[SCN|timestamp expression]|
[before SCN|timestamp expression
```

其中:

standby:指定执行闪回的数据库为备用数据库;

to SCN:将数据库恢复到指定 SCN 状态;

to timestamp:将数据库恢复到指定的时间点;

to before SCN:将数据库恢复到指定 SCN 的前一个 SCN 状态;

to before timestamp:将数据库恢复到指定时间点前 1s 的状态。

13.8 项目案例

视频讲解

13.8.1 闪回查询实例

1. 基于时间戳的闪回查询步骤

除了可以利用时间点 timestamp 或 SCN 查看旧数据外,需要时还可以通过检索旧数据来撤销错误的更改。

步骤1 使用用户 hr 登录，对表 emp 基于 as of timestamp 的闪回查询，其代码如下：

```
SQL> set time on
12:07:48 SQL> conn hr
输入口令：
已连接.
12:07:54 SQL> select employee_id,salary from hr.employees where employee_id = 200;

EMPLOYEE_ID    SALARY
-----------  --------
        200     10000
```

步骤2 更新员工号为 200 的职员工资，更新工资为 2500 和 3000，并提交事务，其代码如下：

```
12:10:36 SQL> update hr.employees set salary = 2500 where employee_id = 200;

已更新 1 行.

12:10:45 SQL> commit;

提交完成.

12:10:48 SQL> update hr.employees set salary = 3000 where employee_id = 200;

已更新 1 行.

12:10:55 SQL> commit;

提交完成.
```

步骤3 更新员工号为 200 的工资，再次更新到 3500 并提交事务，其代码如下：

```
12:11:04 SQL> update hr.employees set salary = 3500 where employee_id = 200;

已更新 1 行.

12:11:27 SQL> commit;

提交完成.
```

步骤4　查看该员工更新后的工资,其代码如下：

```
12:11:29 SQL> select employee_id,salary from hr.employees where employee_id = 200;

EMPLOYEE_ID    SALARY
-----------   -------
        200      3500
```

步骤5　查询该职员前一个小时的工资,其代码如下：

```
12:15:59 SQL> select employee_id,salary from hr.employees as of timestamp sysdate - 1/24 where employee_id = 200;

EMPLOYEE_ID    SALARY
-----------   -------
        200     10000
```

步骤6　查询时间点为"2020-02-03 12:11:10"时该员工的工资,其代码如下：

```
12:19:06 SQL> select employee_id,salary from hr.employees as of timestamp to_timestamp('2020 -
02 - 23 12:11:00','YYYY - MM - DD HH24:MI:SS') where employee_id = 200;

EMPLOYEE_ID    SALARY
-----------   -------
        200      3000
```

步骤7　如果需要,可以将数据恢复到过去某个时刻的状态,其代码如下：

```
12:24:12 SQL> update hr.employees set salary = ( select salary from hr.employees as of timestamp
to_timestamp('2020 - 02 - 23 12:11:00','YYYY - MM - DD HH24:MI:SS') where employee_id = 200)
where employee_id = 200;

已更新 1 行.

12:24:46 SQL> select employee_id,salary from hr.employees where employee_id = 200;

EMPLOYEE_ID    SALARY
-----------   -------
        200      3000
```

2. 基于 SCN 的闪回查询步骤

步骤1　确认 SCN 号。

```
12:24:49 SQL> conn sys as sysdb
输入口令：
已连接.
12:31:32 SQL> select current_SCN from v$database;

CURRENT_SCN
------------
    3554572
```

步骤 2　连续两次更改工资为 5000 和 10000，并分别提交，其代码如下：

```
SQL> update hr.employees set salary = 5000 where  employee_id = 200;
已更新 1 行.

SQL> update hr.employees set salary = 10000 where  employee_id = 200;
已更新 1 行.

SQL> commit;
提交完成.
```

步骤 3　确认更新完成后的 SCN 号，其代码如下：

```
SQL> select current_SCN from v$database;

CURRENT_SCN
------------
    3554707
```

步骤 4　根据 SCN 号读取更新前的工资，其代码如下：

```
SQL> select employee_id,salary from hr.employees as of SCN 3554572 where employee_id = 200;

EMPLOYEE_ID    SALARY
-----------  --------
        200      3000
```

用户根据查询到的结果即可进行数据值的恢复。

注意：事实上，Oracle 内部都是使用 SCN，即使你指定的是 as of timestamp，oracle 也会将其转换成 SCN，系统时间标记与 SCN 之间存在一张表，即 SYS 下的 smon_SCN_time。每隔 5min，系统产生一次系统时间标记与 SCN 的匹配并存入 sys.smon_SCN_time 表。注意理解系统时间标记与 SCN 的每 5min 匹配一次这句话，比如 SCN：339988，339989 分别匹配 08-05-3013：52：00 和 2008-13：57：00，则当你通过 as of timestamp 查询 08-05-30 13：52：00 或 08-05-30 13：56：59 这段时间点内的时间时，oracle 都会将其匹配为 SCN：339988 到

UNDD 表空间中查找,也就说在这个时间内,不管你指定的时间点是什么,查询返回的都将是 08-05-30 13:52:00 这个时刻的数据。

查看 SCN 和 timestamp 之间的对应关系可以使用如下语句:

```
SQL> select SCN,to_char(time_dp,'yyyy-mm-dd hh24:mi:ss') from sys.smon_SCN_time;
```

3. 闪回版本查询的操作步骤

步骤 1 查询两个 SCN 号之间的查询版本,其代码如下:

```
SQL> select versions_xid xid,versions_startSCN startSCN,versions_endSCN
  2  endSCN,versions_operation opera
  3  from hr.employees versions between SCN 3554572 and 3554707
  4  where employee_id = 200 order by startSCN;

XID                STARTSCN   ENDSCN   O
---------------- ---------- -------- 
080014000C070000   3554689            U
                   3554689
```

步骤 2 查询两个时间点之间的查询版本,其代码如下:

```
SQL> col starttime for a30
SQL> col endtime for a30
SQL> select versions_xid xid,versions_starttime starttime,versions_endtime endtime,versions_
operation opera
  2  from hr.employees versions between timestamp
  3  TO_TIMESTAMP('2020-2-23 12:10:36','YYYY-MM-DD HH24:MI:SS') and  TO_TIMESTAMP('2020
     -2-23 12:11:30','YYYY-MM-DD HH24:MI:SS')
  4  where employee_id = 200 order by starttime;

XID              STARTTIME                     ENDTIME                         O
---------------- ----------------------------- -----------------------------
07000A000B070000 23-2月 -20 12.10.45 下午 23-2月 -20 12.10.57 下午   U
05001300FC060000 23-2月 -20 12.10.57 下午                                 U
                                               23-2月 -20 12.10.45 下午
```

在进行版本查询时,在查找过程中会发现如下错误:ORA-30052:下限快照表达式无效。

原因:undo_rentention=900 只记录了 15min,即能上限选择 15min 之前,或者下限选择 15min 之后。

问题解决办法:

修改 undo_retention 的值,具体语句如下所示。

```
SQL> alter system set undo_retention = 10800 scope = both;
```

13.8.2 闪回删除实例

步骤 1 确定回收站状态,其代码如下:

```
SQL> show parameter recyclebin;

NAME                                 TYPE        VALUE
------------------------------------ ----------- ----------
recyclebin                           string      on
```

步骤 2 连接用户 hr,查询当前表,其代码如下:

```
SQL> select * from cat;

TABLE_NAME                           TABLE_TYPE
------------------------------------ ----------
BIN$+vtxfGEtTDiCvoF0vz9Njw==$0       TABLE
T1                                   TABLE
REGIONS                              TABLE
COUNTRIES                            TABLE
LOCATIONS                            TABLE
LOCATIONS_SEQ                        SEQUENCE
DEPARTMENTS                          TABLE
DEPARTMENTS_SEQ                      SEQUENCE
JOBS                                 TABLE
EMPLOYEES                            TABLE
EMPLOYEES_SEQ                        SEQUENCE

TABLE_NAME                           TABLE_TYPE
------------------------------------ ----------
JOB_HISTORY                          TABLE
EMP_DETAILS_VIEW                     VIEW
TEST1                                TABLE
SYS_TEMP_FBT                         TABLE
COURSE                               TABLE
DEPT_DUMP                            TABLE
BIN$uFQDoXrNRq+yKLGuPuqoIQ==$0       TABLE
BIN$EG5F0qHRRC2D+LTe0Gr+fA==$0       TABLE
EMP_DUMP                             TABLE

已选择 20 行.
```

步骤3　模拟误删除表 dept_dump，并查询删除后的变化，其代码如下：

```
SQL> drop table dept_dump;
表已删除.

SQL> select * from cat;

TABLE_NAME                                TABLE_TYPE
----------------------------------------  ----------
BIN$+vtxfGEtTDiCvoF0vz9Njw==$0            TABLE
BIN$rn9H+MrKQkGgl4zhq3vpuA==$0            TABLE
T1                                        TABLE
REGIONS                                   TABLE
COUNTRIES                                 TABLE
LOCATIONS                                 TABLE
LOCATIONS_SEQ                             SEQUENCE
DEPARTMENTS                               TABLE
DEPARTMENTS_SEQ                           SEQUENCE
JOBS                                      TABLE
EMPLOYEES                                 TABLE
EMPLOYEES_SEQ                             SEQUENCE
JOB_HISTORY                               TABLE
EMP_DETAILS_VIEW                          VIEW
TEST1                                     TABLE
SYS_TEMP_FBT                              TABLE
COURSE                                    TABLE
BIN$uFQDoXrNRq+yKLGuPuqoIQ==$0            TABLE
BIN$EG5F0qHRRC2D+LTe0Gr+fA==$0            TABLE
EMP_DUMP                                  TABLE

已选择 20 行.
```

步骤4　查询回收站，其代码如下：

```
SQL> select object_name,ts_name from user_recyclebin;

OBJECT_NAME                               TS_NAME
----------------------------------------  -------
BIN$+vtxfGEtTDiCvoF0vz9Njw==$0            SYSAUX
BIN$uFQDoXrNRq+yKLGuPuqoIQ==$0            SYSAUX
BIN$EG5F0qHRRC2D+LTe0Gr+fA==$0            SYSAUX
BIN$rn9H+MrKQkGgl4zhq3vpuA==$0            SYSAUX
```

```
SQL> show recyclebin;
ORIGINAL NAME    RECYCLEBIN NAME                              OBJECT TYPE   DROP TIME
-------------    ---------------------------------------      -----------   -------------------
DEPT_DUMP        BIN$ rn9H+MrKQkGgl4zhq3vpuA==$0              TABLE         2020-02-23:13:02:23
EMP_DUMP         BIN$ EG5F0qHRRC2D+LTe0Gr+fA==$0              TABLE         2020-02-21:16:59:25
EMP_DUMP         BIN$ uFQDoXrNRq+yKLGuPuqoIQ==$0              TABLE         2020-02-21:12:57:43
EMP_DUMP         BIN$ +vtxfGEtTDiCvoF0vz9Njw==$0              TABLE         2019-12-12:13:38:52
```

步骤 5 闪回删除，其代码如下：

```
SQL> flashback table dept_dump to before drop;
```

闪回完成.

步骤 6 观察闪回后的回收站以及表 cat 信息，其代码如下：

```
SQL> show recyclebin;
ORIGINAL NAME    RECYCLEBIN NAME                              OBJECT TYPE   DROP TIME
-------------    ---------------------------------------      -----------   -------------------
EMP_DUMP         BIN$ EG5F0qHRRC2D+LTe0Gr+fA==$0              TABLE         2020-02-21:16:59:25
EMP_DUMP         BIN$ uFQDoXrNRq+yKLGuPuqoIQ==$0              TABLE         2020-02-21:12:57:43
EMP_DUMP         BIN$ +vtxfGEtTDiCvoF0vz9Njw==$0              TABLE         2019-12-12:13:38:52
SQL> select * from cat;

TABLE_NAME                              TABLE_TYPE
------------------------------          ------------
BIN$ +vtxfGEtTDiCvoF0vz9Njw==$0         TABLE
DEPT_DUMP                               TABLE
T1                                      TABLE
REGIONS                                 TABLE
COUNTRIES                               TABLE
LOCATIONS                               TABLE
LOCATIONS_SEQ                           SEQUENCE
DEPARTMENTS                             TABLE
DEPARTMENTS_SEQ                         SEQUENCE
JOBS                                    TABLE
EMPLOYEES                               TABLE
EMPLOYEES_SEQ                           SEQUENCE
JOB_HISTORY                             TABLE
EMP_DETAILS_VIEW                        VIEW
TEST1                                   TABLE
SYS_TEMP_FBT                            TABLE
```

```
COURSE                                          TABLE
BIN $ uFQDoXrNRq + yKLGuPuqoIQ = = $ 0          TABLE
BIN $ EG5F0qHRRC2D + LTe0Gr + fA = = $ 0        TABLE
EMP_DUMP                                        TABLE
```

已选择 20 行.

步骤 7　演示 purge 选项的删除效果,其代码如下:

```
SQL> drop table dept_dump purge;
```

等同于

```
SQL> drop table dept_dump;
SQL> purge recyclebin;
```

然后观察使用 purge 后的删除效果,其代码如下:

```
SQL> drop table dept_dump purge;

表已删除.

SQL> show recyclebin;
ORIGINAL NAME    RECYCLEBIN NAME                          OBJECT TYPE   DROP TIME
---------------  ---------------------------------------  -----------   -------------------
EMP_DUMP         BIN $ EG5F0qHRRC2D + LTe0Gr + fA = = $ 0  TABLE        2020 - 02 - 21:16:59:25
EMP_DUMP         BIN $ uFQDoXrNRq + yKLGuPuqoIQ = = $ 0    TABLE        2020 - 02 - 21:12:57:43
EMP_DUMP         BIN $ + vtxfGEtTDiCvoF0vz9Njw = = $ 0     TABLE        2019 - 12 - 12:13:38:52
SQL> select * from cat;

TABLE_NAME                              TABLE_TYPE
------------------------------------    ----------
BIN $ + vtxfGEtTDiCvoF0vz9Njw = = $ 0   TABLE
T1                                      TABLE
REGIONS                                 TABLE
COUNTRIES                               TABLE
LOCATIONS                               TABLE
LOCATIONS_SEQ                           SEQUENCE
DEPARTMENTS                             TABLE
DEPARTMENTS_SEQ                         SEQUENCE
JOBS                                    TABLE
EMPLOYEES                               TABLE
EMPLOYEES_SEQ                           SEQUENCE
JOB_HISTORY                             TABLE
```

```
EMP_DETAILS_VIEW                          VIEW
TEST1                                     TABLE
SYS_TEMP_FBT                              TABLE
COURSE                                    TABLE
BIN $ uFQDoXrNRq + yKLGuPuqoIQ = = $ 0    TABLE
BIN $ EG5F0qHRRC2D + LTe0Gr + fA = = $ 0  TABLE
EMP_DUMP                                  TABLE

已选择 19 行.
```

13.8.3 闪回表实例

视频讲解

步骤1 使用用户 hr 登录,查询表 tab,其代码如下:

```
SQL> select * from tab;

TNAME                                     TABTYPE        CLUSTERID
--------------------------------------    ------------   ----------
BIN $ + vtxfGEtTDiCvoF0vz9Njw = = $ 0    TABLE
T1                                        TABLE
REGIONS                                   TABLE
COUNTRIES                                 TABLE
LOCATIONS                                 TABLE
DEPARTMENTS                               TABLE
JOBS                                      TABLE
EMPLOYEES                                 TABLE
JOB_HISTORY                               TABLE
EMP_DETAILS_VIEW                          VIEW
TEST1                                     TABLE
SYS_TEMP_FBT                              TABLE
COURSE                                    TABLE
BIN $ uFQDoXrNRq + yKLGuPuqoIQ = = $ 0    TABLE
BIN $ EG5F0qHRRC2D + LTe0Gr + fA = = $ 0  TABLE
EMP_DUMP                                  TABLE

已选择 16 行.
```

步骤2 创建表 test2,插入记录,提交事务,其代码如下:

```
SQL> create table test2(id int,name varchar2(10));
表已创建.
```

```
SQL> insert into test2 values(1,'stu1');
已创建 1 行.

SQL> insert into test2 values(2,'stu2');
已创建 1 行.

SQL> insert into test2 values(3,'stu3');
已创建 1 行.

SQL> commit;
提交完成.
```

步骤3 查询当前 SCN 号。如果当前用户没有权限查询 v$database,则以用户 sys 登录,并授予当前用户访问数据字典的权限,其代码如下:

```
SQL> conn sys as sysdb
输入口令:
已连接.
SQL> grant select any dictionary to hr;

授权成功.

SQL> conn hr
输入口令:
已连接.
SQL> select current_SCN from v$database;

CURRENT_SCN
-----------
    3558225
```

步骤4 更新记录,并提交事务,其代码如下:

```
SQL> update test2 set name = 'stu**' where id = 1;

已更新 1 行.

SQL> commit;

提交完成.
```

步骤5 看表中的记录,其代码如下:

```
SQL> select * from test2;

        ID  NAME
---------- ----
         1  stu**
         2  stu2
         3  stu3
```

步骤 6 删除 id=3 的记录，其代码如下：

```
SQL> delete from test2 where id = 3;

已删除 1 行.

SQL> commit;

提交完成.
```

步骤 7 启动表 test2 的 row movement 特性，其代码如下：

```
SQL> alter table test2 enable row movement;

表已更改.
```

步骤 8 将表 test2 恢复到 SCN 为 3558225 的状态，其代码如下：

```
SQL> flashback table test2 to SCN 3558225;

闪回完成.
SQL> select * from test2;

        ID  NAME
---------- ----
         1  stu1
         2  stu2
         3  stu3
```

13.8.4 闪回数据库实例

视频讲解

步骤 1 确定闪回区的设置。
（1）确认数据库处于归档模式，其代码如下：

```
SQL> conn sys as sysdb
输入口令:
已连接.
SQL> archive log list
数据库日志模式              存档模式
自动存档                   启用
存档终点                   d:\backup\
最早的联机日志序列           1
下一个存档日志序列           2
当前日志序列                2
```

(2) 数据库是否启用闪回数据库,其代码如下:

```
SQL> select flashback_on from v$database;

FLASHBACK_ON
-------------
YES
```

(3) 查看闪回区相应参数,其代码如下:

```
SQL> show parameter db_flashback_retention_target;

NAME                                 TYPE        VALUE
------------------------------------ ----------- ------------------------------
db_flashback_retention_target        integer     1440
SQL> show parameter db_recovery_file;

NAME                                 TYPE        VALUE
------------------------------------ ----------- ------------------------------
db_recovery_file_dest                string      D:\Oracle19c\fast_recovery_area\
db_recovery_file_dest_size           big integer 500M
```

注意:如果关闭了闪回数据库的功能,就使用 SQL> alter database flashback on;开启。

步骤 2 使用 SCN 闪回数据库。

(1) 查看数据库系统的当前 SCN,其代码如下:

```
SQL> set time on
13:32:32 SQL> select current_SCN from v$database;

CURRENT_SCN
-------------
    3558931
```

(2) 改变数据库当前状态,模拟创建表 test3,并插入一条记录,其代码如下:

```
13:32:36 SQL> create table test3(id int,name varchar2(10));

表已创建.

13:33:36 SQL> insert into test3 values(1,'stu1');

已创建 1 行.

13:33:41 SQL> commit;

提交完成.
```

(3) 进行闪回数据库恢复,将数据库恢复到创建表之前的状态,即 SCN 为 3558931,其代码如下:

```
13:33:43 SQL> shutdown immediate;
数据库已经关闭.
已经卸载数据库.
ORACLE 例程已经关闭.
13:35:52 SQL> startup mount;
ORACLE 例程已经启动.

Total System Global Area 2550136752 bytes
Fixed Size                   9031600 bytes
Variable Size             1191182336 bytes
Database Buffers          1342177280 bytes
Redo Buffers                 7745536 bytes
数据库装载完毕.
13:36:55 SQL> flashback database to SCN 3558931;

闪回完成.
```

(4) 用 resetlogs 选项打开数据库,其代码如下:

```
13:37:08 SQL> alter database open resetlogs;

数据库已更改.
```

(5) 验证数据库的状态,其代码如下:

```
13:37:48 SQL>select * from test3;
select * from test3
              *
第 1 行出现错误:
ORA-00942: 表或视图不存在
```

结果发现,表 test3 不存在。

步骤 3 按照指定时间闪回数据库。

(1) 查询数据库中当前最早的闪回 SCN 和时间,其代码如下:

```
13:38:51 SQL>alter session set nls_date_format = 'yyyy-mm-dd hh24:mi:ss';

会话已更改.

13:40:53   SQL>select   oldest_flashback_SCN,oldest_flashback_time   from v$flashback_
database_log;

OLDEST_FLASHBACK_SCN OLDEST_FLASHBACK_TI
-------------------- -------------------
             3549607 2020-02-23 11:58:38
```

(2) 查询数据库中当前时间和当前 SCN,其代码如下:

```
13:41:24 SQL>select sysdate from dual;

SYSDATE
-------------------
2020-02-23 13:41:53
```

(3) 改变数据库的当前状态,模拟创建表 test4,并插入 1 条记录,其代码如下:

```
13:41:53 SQL>create table test4(id int,name varchar2(20));

表已创建.

13:43:29 SQL>insert into test4 values(1,'stu1');

已创建 1 行.

13:43:35 SQL>commit;

提交完成.
```

(4)进行闪回数据库恢复,将数据库恢复到创建表之前的状态,其代码如下:

```
13:43:37 SQL> shutdown immediate;
数据库已经关闭.
已经卸载数据库.
ORACLE 例程已经关闭.
13:45:05 SQL> startup mount;
ORACLE 例程已经启动.

Total System Global Area  2550136752 bytes
Fixed Size                   9031600 bytes
Variable Size             1191182336 bytes
Database Buffers          1342177280 bytes
Redo Buffers                 7745536 bytes
数据库装载完毕.
13:46:21 SQL> flashback database to timestamp to_timestamp('2020-02-23 13:41:53','YYYY-MM
-DD HH24:MI:SS');

闪回完成.
```

(5)打开数据库,其代码如下:

```
13:46:37 SQL> alter database open resetlogs;

数据库已更改.
```

(6)验证数据库的状态,以查看表 test4 是否存在,其代码如下:

```
13:47:55 SQL> select * from test4;
select * from test4
              *
第 1 行出现错误:
ORA-00942: 表或视图不存在
```

闪回数据库操作的限制:

(1)当发生数据文件损坏或丢失等介质故障时,不能使用闪回数据库进行恢复。闪回数据库只能基于当前正常运行的数据文件。

(2)在闪回数据库功能启动后,如果发生数据控制文件重建或利用备份恢复控制文件,则不能使用闪回数据库。

(3)不能使用闪回数据库进行数据文件收缩操作。

(4)不能使用闪回数据库将数据库恢复到在闪回日志中可获得的最早的 SCN 之前的 SCN,因为在一定的条件下被删除的闪回日志文件,不是始终保存在闪回恢复区中的。

第14章

Oracle并发与一致性

PPT 视频讲解

14.1 并发性与一致性

1. 并发性与一致性的概念

在单用户的数据库中,用户可以修改数据,而不用担心其他用户在同一时间修改相同的数据。但是在一个多用户的数据库中,多个事务内的语句可以同时更新相同的数据。注意,同时执行的多个事务必须产生有意义且一致的结果。因此,多用户数据库必须提供以下两个功能。

数据并发性:确保多个用户可以同时访问数据。

数据一致性:确保每个用户看到数据的一致的视图,包括可以看到用户自己的事务所做的更改和其他用户已提交的事务所做的更改。

2. 不一致性问题

并发会产生多个事务同时存取同一数据的情况。可能存取和存储不正确的数据,破坏事务的一致性和数据库的一致性。归纳起来共有以下 3 个主要问题。

(1) 脏读:在事务中,读到了其他事务没有提交的记录,如图 14-1 所示。

时间点	任务1	任务2
1	X=R(A)=100	
2	X+=50	
3	W(A, X) = 150	
4		X=R(A)=150
5	ROLLBACK	
6		X+=40
7		W(A, X) = 190
8		COMMIT
串行执行结果:140		并发执行结果:190

图 14-1 脏读现象

[分析]

串行化结果：x:=140。

并型：R1(a),W1(A),R2(A),ROLLBACK1,W2(A),COMMIT2。

结果：x:=190,错误。

问题：第 4 步,A2 读了 A1 写后未提交的脏数据。

(2) 不可重复读：在事务中,即使查询条件相同,下次返回的记录与上次返回的记录也不一样。其不一样体现在记录被修改或记录被删除,如图 14-2 所示。

时间点	任务1	任务2
1	X=R(A)=100	
2		X=R(A)=100
3	X+=50	
4	W(A, X) = 150	
5	COMMIT	
6		X+=40
7		W(A, X) = 140
8		COMMIT

串行执行结果：190	并发执行结果：140

图 14-2 不可重复读

[分析]

串行化结果：X:=190。

并行：R1(A),R2(A),S1(A),COMMIT,W2(A),COMMIT2。

结果：X:=140,错误。

问题：第 2,6 时间点,执行第 6 时间点前 X 已经不是第 2 时间点读出的 100,已经是 140 了。A1 对 A2 正在使用的记录数据做了修改。

区别：第 2 时间点中 A2 没有读 A1 修改但尚未提交的脏数据(不是脏读)；第 7 时间点中 A2 没有写 A1 修改且尚未提交的脏数据(不是脏写)。

(3) 幻影读：在事务中,即使查询条件相同,下次返回的记录与上次返回的记录也不一样。其不一样体现在新增加了记录,如图 14-3 所示。

时间点	任务1	任务2
	原有4行数据, 总和为1000	
1	S=SELECT SUM(A) ... = 1000	
2		INSERT (A) VALUES (200)
3		COMMIT
4	N=SELECT COUNT(*) ... =5	
5	AVG=S/N = 200	
6		

串行执行结果：250或240	并发执行结果：200

图 14-3 幻影读

[分析]

串行化结果:两个事务都满足规则,则返回 YES。

并行:同一事物中相同的两次查询结果都一样,结果集记录的个数却增加一个。

Oracle 中,使用谓词锁避免幻影读,即满足 select 语句中 where 谓词的记录都不能被 insert,其他可以。

3. 解决办法

要想解决脏读、不可重复读、幻影读等不一致的问题,就需要提高事务的隔离级别。

为了描述事务并发运行时的一致性行为,研究人员定义了一种事务隔离模型,称之为序列化(Serializability)。这种可串行化事务操作使得它看起来似乎没有其他用户在操作数据。虽然这种序列化机制在一般情况下是可用的,但在并发要求高的场景,它会严重影响系统的吞吐能力,即事务的隔离级别越高,并发能力也就越低。所以,一般情况下,需要在事务隔离级别与性能间进行取舍。

从运维的角度来看,提高事务隔离级别的解决办法可以总结如下所述。

(1) 在通常情况下,程序员无须考虑每个事务内部的处理细节来避免并发异常,而是通过控制隔离级别来保证不出现相关的异常。

(2) 程序员可以采用以下两种办法尽量避免可能的并发错误。

① 使用尽量少的 SQL,直接修改数据。

② 可以使用手动加锁等方式预先申请资源。

14.2 隔离机制

为了兼顾并发效率和异常控制,在标准 SQL 规范中,定义了如图 14-4 所示的 4 个事务隔离级别以及各隔离级别对各种异常的控制能力。

隔离性	Isolation Level	Dirty Read	Non Repeatable Read	Phantom Read	性能
↓	Read Uncommitted	Possible	P	P	↑
	Read Committed	Not	P	P	
	Repeatable Read	Not	Not	P	
	Serializable	Not	Not	Not	

图 14-4 事务隔离级别

Oracle 数据库提供了 Read Committed(默认级别)和 Serializable 两种隔离级别,同时还支持只读模式。

1. Read Committed 事务隔离级别

在此级别中,事务中查询到的数据都是在此查询前已经提交的。这种隔离级别避免了读取脏数据。然而数据库并不阻止其他事务修改一个所读取的数据,其他事务可能在查询执行期间修改。因此,一个事务运行同样的查询两次,可能发生不可重复读和幻影读的情况。

2. Serializable 事务隔离级别

简单地说，Serializable 就是使事务看起来像是一个接着一个地顺序地执行。仅仅能看见在本事务开始前由其他事务提交的更改和在本事务中所做的更改。保证不会出现非重复读和幻象。Serializable 隔离级别提供了 Read-Only 事务所提供的读一致性（事务级的读一致性），同时又允许执行数据操纵类型的操作。

3. Read-Only 事务隔离级别

只读隔离级别与序列化隔离级别很像，只是在只读事务中，除了用户 sys 外，不允许有修改操作。因此只读事务不会有 ORA-08177 错误，只读事务在产生一个报告时很有效。

14.3 锁机制

事务之间的并发控制实际是通过锁实现的，锁是用来预防事务之间访问相同数据时的破坏性交互（如错误的更新数据等）的一种机制，在维护数据库并发性与一致性方面扮演了一个重要角色。

1. 自动锁

Oracle 数据库在执行 SQL 的时候会自动获取所需要的锁，因此用户在应用设计的时候只需要定义恰当的事务级别，而不需要显示锁定任何资源（即使 Oracle 提供了手动锁定数据的方法，用户不会用到）。

2. Oracle 锁模式

Oracle 自动使用最低的限制级别去提供最高程度的并发。Oracle 中有以下两种类型的锁。

（1）排他锁（X）：这种类型是阻止资源共享的，一个事务获取了一个排他锁，则这个事务锁定期间只有这个事务可以修改这个资源。

（2）共享锁（S）：这种类型是允许资源共享的，多个事务可以在同一资源上获取共享锁。

假设一个事务使用 select…for update（其他 DML 操作也一样）来选择表中的一行，则这个事务会获取该行的排他行锁以及所在表的共享表锁。行锁允许其他事务修改除该行的其他行，而表锁阻止其他事务修改表结构。

3. Oracle 封锁粒度

DML 锁可以防止多个相互冲突的 DDL 和 DML 操作对数据的破坏性。DML 语句自动获取两种类型的锁：Row locks(TX) 和 Table locks(TM)。封锁粒度如图 14-5 所示。

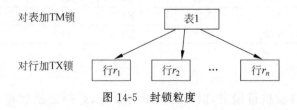

图 14-5 封锁粒度

4. 意向锁

表是由行组成的，当向某个表加锁时，一方面需要检查该锁的申请是否与原有的表级锁相容；另一方面，还要检查该锁是否与表中的每行上的锁相容。比如一个事务要在一个表上加 S 锁，如果表中的一行已被另外的事务加了 X 锁，那么该锁的申请也应被阻塞。如果表中的数据很多，逐行检查锁标志的工作量将会很大，系统的性能也会受到影响。为了解决这个问题，可以在表级引入新的锁类型来表示其所属行的加锁情况，于是引出了"意向锁"的概念。

意向锁的含义是如果对一个节点加意向锁，则说明该节点的下层节点正在被加锁。当对任一节点加锁时，必须先对它的上层节点加意向锁。例如对表中的任一行加锁时，必须先对它所在的表加意向锁，然后再对该行加锁。这样一来，当事务对表加锁时，就不再需要检查表中每行记录的锁标志位了，系统效率得以大大提高。

锁有 S 锁和 X 锁两种基本的锁类型，可以自然地派生出以两种意向锁。

(1) 意向共享锁(Intent Share Lock, IS 锁)：如果要对一个数据库对象加 S 锁，首先要对其上级节点加 IS 锁，表示它的后裔节点拟(意向)加 S 锁。

(2) 意向排它锁(Intent Exclusive Lock, IX 锁)：如果要对一个数据库对象加 X 锁，首先要对其上级节点加 IX 锁，表示它的后裔节点拟(意向)加 X 锁。

另外，基本的锁类型(S、X)与意向锁类型(IS、IX)之间还可以组合出新的锁类型，理论上可以组合出 4 种，即 S+IS, S+IX, X+IS, X+IX。但稍加分析不难看出，实际上只有 S+IX 有新的意义，其他 3 种组合都没有使锁的强度得到提高(即 S+IS=S, X+IS=X, X+IX=X，这里的"="指锁的强度相同)。所谓锁的强度是指对其他锁的排斥程度。这样又可以引入一种新的锁的类型——共享意向排它锁(SharedIntent Exclusive Lock, SIX 锁)。如果对一个数据库对象加 SIX 锁，表示对它加 S 锁，再加 IX 锁，即 SIX=S+IX。例如，若事务对某个表加 SIX 锁，则表示该事务要读整个表(所以要对该表加 S 锁)，同时会更新个别行(所以要对该表加 IX 锁)。这样数据库对象上所加的锁类型就可能有 5 种，即 S、X、IS、IX、SIX。

5. TM 锁的相容矩阵

TM 锁的相容矩阵如表 14-1 所示。

表 14-1　TM 锁的相容矩阵

事务 B ＼ 事务 A	S	X	RS	RX	SRX	—
S	YES	NO	YES	NO	NO	YES
X	NO	NO	NO	NO	NO	YES
RS	YES	NO	YES	YES	YES	YES
RX	NO	NO	YES	YES	YES	YES
SRX	NO	NO	YES	NO	NO	YES
—	YES	YES	YES	YES	YES	YES

6. 手工锁

Oracle 数据库自动执行锁定，以确保数据并发性、数据完整性和与数据读取一致性。但是也可以使用手动锁覆盖默认的锁机制。例如，事务中包含如下语句时会覆盖 Oracle 的

默认锁:

```
set transaction isolation level
lock table,
select... for update
```

14.4 项目案例

14.4.1 Read Committed 隔离级别实例

事务1(可以是 Read Committed 或 Serializable)与事务2(Read Committed)的典型交互,称为 Lost Update(丢失更新),具体如表14-2操作步骤所示。

表14-2 Read Committed 隔离级别实例操作步骤

事务1	事务2	说明
SQL> select last_name, salary from employees where last_name in ('Banda','Greene','Hintz'); LAST_NAME SALARY ――――――――― Banda 6200 Greene 9500		
SQL> update employees set salary = 7000 where last_name = 'Banda';		事务1用的是默认隔离级别 read committed
	SQL> set transaction isolation level read committed;	
	SQL> select last_name, salary from employees where last_name in ('Banda','Greene','Hintz'); LAST_NAME SALARY ――――――――― Banda 6200 Greene 9500	事务2通过使用 Oracle 的读一致性得到了事务1更新前的数据
	SQL> update employees set salary = 9900 where last_name = 'Greene';	
SQL> insere into employees (employee_id, last_name, email, hire_date, job_id) value (210, 'Hintz', 'JHINTZ', sysdate, 'SH_CLERK');		事务1插入了 employee Hintz,但并没有提交

续表

事务 1	事务 2	说明
	SQL> select last_name, salary from employees where last_name in ('Banda','Greene','Hintz'); LAST_NAME　　　SALARY Banda　　　　　6200 Greene　　　　　9500	事务 2 看不到事务 1 未提交的 Hintz 信息
	SQL> update employees set salary = 6300 where last_name = 'Banda'; —— prompt does not return	事务 2 尝试去更新被事务锁住的 Banda 信息,产生了写冲突,此时事务 2 要等到事务 1 结束后再执行
SQL> commit;		
	1 row updated. SQL>	事务 1 提交,结束了事务,事务 2 继续处理
	SQL> select last_name, salary from employees where last_name in ('Banda','Greene','Hintz'); LAST_NAME　　　SALARY Banda　　　　　6300 Greene　　　　　9900 Hintz COMMIT;	
SQL> select last_name, salary from employees where last_name in ('Banda','Greene','Hintz'); LAST_NAME　　　SALARY Banda　　　　　6300 Greene　　　　　9900 Hintz		

14.4.2　Serializable 事务隔离级别实例

表 14-3 所示的是一个序列化事务是如何与其他事务交互的。如果一个序列化任务不去尝

视频讲解

试修改其他事务在序列化事务开始后提交的数据,那么 Serialized Access 问题就可以避免了。

表 14-3 序列化事务隔离级别实例操作步骤

事务 1	事务 2	说明
SQL> select last_name, salary from employees where last_name IN ('Banda','Greene','Hintz'); LAST_NAME　　　SALARY Banda　　　　　　6200 Greene　　　　　　9500		
SQL> update employees set salary = 7000 where last_name = 'Banda';		事务 1 是默认的 read committed
	SQL> set transaction isolation level serializable;	
	SQL> select last_name, salary from employees where last_name IN ('Banda','Greene','Hintz'); LAST_NAME　　　SALARY Banda　　　　　　6200 Greene　　　　　　9500	
	SQL> update employees set salary = 9900 where last_name = 'Greene';	
SQL> insert into employees (employee_id, last_name, email, hire_date, job_id) values (210, 'Hintz', 'JHINTZ', SYSDATE, 'SH_CLERK');		
SQL> COMMIT;		
SQL> select last_name, salary from employees where last_name IN ('Banda','Greene','Hintz'); LAST_NAME　　　SALARY Banda　　　　　　7000 Greene　　　　　　9500 Hintz	SQL> select last_name, salary from employees where last_name in ('Banda','Greene','Hintz'); LAST_NAME　　　SALARY Banda　　　　　　6200 Greene　　　　　　9900	注意,Oracle 的读一致性使得事务 2 的前后读取是一致的,即事务 1 的插入和更新操作对事务 2 来说是不可见的
	COMMIT;	

续表

事务 1	事务 2	说明
SQL> select last_name, salary from employees where last_name IN ('Banda','Greene','Hintz'); LAST_NAME SALARY ——————————————— Banda 7000 Greene 9900 Hintz	SQL> select last_name, salary from employees where last_name IN ('Banda','Greene','Hintz'); LAST_NAME SALARY ——————————————— Banda 7000 Greene 9900 Hintz	
SQL> update employees set salary = 7100 where last_name = 'Hintz';		
	SQL> set transaction isolation level serializable;	
	SQL> update employees set salary = 7200 where last_name = 'Hintz'; —— prompt does not return	
SQL> COMMIT;		
	update employees set salary = 7200 where last_name = 'Hintz' * error at line 1: ORA-08177: can't serialize access for this transaction	报错原因在于事务 3 的提交是在事务 4 开始之后,没有满足序列化
	SQL> rollback;	事务 2 回滚以结束事务

14.4.3 提交读与序列实例

1. 提交读

视频讲解

这是 Oracle 默认的事务隔离级别。Oracle 中任何事务中的任何一条语句都必须遵从语句级的读一致性。在这种一致性的前提下,保证不会脏读,但可出现非重复读和幻象。从语句和事务两个层次分析,提交读具有下面两个特点。

（1）语句开始执行时拍一个快照,在该语句的执行过程中,所有的数据以快照为准。

（2）在事务处理过程中,一个事务不能看到另外一个未提交事务的修改。

在语句执行过程中,看不到别的事务的任何修改,而对于"事务处理过程中,一个事务不能看到另外一个事务的未提交的修改",通过实例来说明这个问题,如表 14-4 所示。

第14章 Oracle并发与一致性

表 14-4 提交读实训操作步骤

事务 A	事务 B
SQL> select * from test; No rows selected	
	SQL> insert into test values(1) 1 row created. SQL> insert into test values(2) 1 row created. SQL> insert into test values(3) 1 row created.
SQL> select * from test; No rows selected	
	SQL> commit Commit complete.
SQL Select * from test; 　　A ――― 　　1 　　2 　　3	

2. 序列化

序列化(Serializable)可以使事务看起来像是一个接着一个地顺序地执行。事务在开始前对数据库中的所有数据拍一个快照,在事务执行过程中仅能看到这个快照中的数据(仅仅能看见在本事务开始前由其他事务提交的更改)和在本事务中所做的更改。表 14-5 所示的是在设置序列化的隔离级别后,即使 A 事务提交,B 事务仍然看不到 A 事务对数据的修改。

表 14-5 序列化隔离级别实例操作步骤

事务 A	事务 B
SQL> set transaction isolation level serializable; Transaction set SQL> select * from test no rows selected	
	SQL> set transaction isolation level serializable; Transaction set SQL> insert into test values(1); 1 rows created SQL> insert into test values(2); 1 rows created SQL> insert into test values(3); 1 rows created

续表

事务 A	事务 B
	SQL> insert into test values(4);
	1 rows created
	SQL> commit
	Commit completed
SQL> select * from test; No rows selected	

14.4.4 行锁定实例

在下面的更新语句中,email 和 phone_number 都是原始的值,这样更新,能够避免上面例子中提到的丢失更新的问题。表 14-6 所示的是当两个会话在相同时间更新 employees 表中相同的行时的执行顺序。

表 14-6 两个会话在相同时间更新 employees 表中相同的行时的执行顺序

时间	会议 1	会议 2	说明
t0	select employee_id, email, phone_number from hr.employees where last_name = 'Himuro'; EMPLOYEE_ID EMAIL PHONE_NUMBER ——————— ——————— 118 GHIMURO 515.127.4565		
t1		select employee_id, email, phone_number from hr.employees where last_name = 'Himuro'; EMPLOYEE_ID EMAIL PHONE_NUMBER ——————— ——————— 118 GHIMURO 515.127.4565	
t2	update hr.employees set phone_number='515.555.1234' where employee_id=118 and email='GHIMURO' and phone_number='515.127.4565'; 1 row updated.		会话 1 获得了修改行的行锁

续表

时间	会议 1	会议 2	说明
t3		update hr.employees set phone_number='515.555.1235' where employee_id=118 and email='GHIMURO' and phone_number='515.127.4565'; —— SQL Plus does not show —— a row updated message or —— return the prompt.	会话 2 尝试获取相同行的行锁，但发生了阻塞
t4	commit; Commit complete.		会话 1 提交
t5		0 rows updated.	因为 where 条件不匹配，所以会话 2 没有更新任何记录
t6	update hr.employees set phone_number='515.555.1235' where employee_id=118 and email='GHIMURO' and phone_number='515.555.1234'; 1 row updated.		
t7		select employee_id, email, phone_number from hr.employees where last_name = 'Himuro'; EMPLOYEE_ID EMAIL PHONE_NUMBER ——————— ——— 118 GHIMURO 515.555.1234	Oracle 的读一致性使得会话 2 看不到未提交的 t6 的修改
t8		update hr.employees set phone_number='515.555.1235' where employee_id=118 and email='GHIMURO' and phone_number='515.555.1234'; —— SQL Plus does not show —— a row updated message or —— return the prompt.	

续表

时间	会议 1	会议 2	说明
t9	rollback; Rollback complete.		
t10		1 row updated.	会话 2 中发现一行匹配的记录
t11		commit; Commit complete.	

Oracle 数据库在执行 SQL 时会自动获取所需要的锁，因此用户在应用设计时只需要定义恰当的事务级别即可，不需要显示锁定任何资源。

第15章

Oracle优化

PPT 视频讲解

15.1 Oracle 性能优化概述

一个设计良好的数据库系统很少遇到效率问题，即一般不需要经常优化。通过大量数据库系统所进行的调查结果显示，绝大多数系统的效率问题都是由于数据库系统的设计和程序的开发所引起的。因此，如果数据库系统的设计和应用程序的开发阶段的质量能够得到足够的保证，多数数据库系统的优化工作都会变得十分轻松。一个好的数据库系统，设计固然是关键，但当软件开发完成后，通过不断地对数据库系统进行跟踪监控和优化也是必不可少的。可以认为，优化是数据库设计的一种顺延，与设计是相辅相成的。

1. 性能调整需要的角色分配

Oracle 性能优化涉及的方面很多，贯穿整个数据库管理系统始末，需要各个角色的数据库从业者通力配合，而单纯依靠数据库管理员（DBA）只能完成部分优化功能。Oracle 性能调整需要的角色分配如图 15-1 所示。

业务分析人员	调整业务功能
设计人员	调整数据库设计 调整流程设计
程序开发人员	调整SQL语句 调整物理结构
数据库管理员	调整内存分配 调整I/O 调整内存竞争
系统管理员	调整操作系统

图 15-1 Oracle 性能调整需要的角色分配

2. 应用周期中调整的代价

多数用户认为,只有在性能差时才需进行调优。但这时进行调优会使某些最有效的调整策略发挥不了作用。如果用户不愿意重新设计应用,则只能通过内存优化和 I/O 优化的办法提高性能。

通过良好的系统设计,就可以在应用系统的生命周期中消除性能调整的代价和挫折。图 15-2 所示的是在应用系统的生命周期中,调整的相对代价和收益。由图可知,最有效的调整时间是在设计阶段,在设计阶段进行的调整,能以最低的代价获得最大的收益。

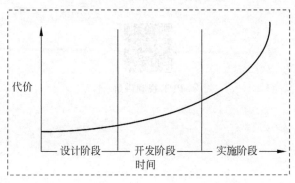

图 15-2 应用周期中调整的代价示意

对于有明显提示的错误,只要时间充足,大多数 DBA 都能通过错误提示对未知问题做出基本的判断,找到改正错误的方向。但是在大多数数据库操作人员的眼里,数据库的性能优化是看不见摸不着、难以掌握的。之所以带给人这样的感觉,是因为数据库在出现性能问题时不报错,这就给诊断问题、解决问题带来了一定的难度。

3. 数据库性能优化重点

在实际运作中,最好的切入点是优化数据库系统内存。就像本书第 1 章介绍的数据库体系结构中讲到的,只要将大多数磁盘(I/O)操作转换成内存操作,数据库系统的效率就会显著提高。之后就要考虑优化数据库系统的 I/O,因为如果能减少 I/O 的数据量或平衡 I/O,系统的效率也会大大提高。这两个部分将是 DBA 继续优化工作的重点。

4. 数据库性能优化目标

数据库性能优化并没有统一的标准步骤,"不管黑猫白猫,捉住老鼠就是好猫"这句话在数据库性能优化领域也同样适用。一个性能优化项目的目标可以简述如下。

(1) 增大数据库吞吐量。如通过数据库优化增加每秒执行的事务数。

(2) 高效利用服务器硬件资源。如开启并充分利用 CPU 和存储 I/O 资源,以加快执行速度。

5. 数据库性能优化基本方法

尽管引起数据库性能故障的原因千变万化,但是解决方法还是有一定规律可循的。下面介绍一些常见的性能优化的基本方法。

(1) 建立数据库性能基线(Baseline)。观察数据库运行正常时的各种指标,主要包括主机资源和数据库资源的消耗情况。

(2) 寻找关键变化,对照性能基线找出偏差所在。

（3）定位主要影响因素，在进行数据库性能优化时抓住主要矛盾，找出影响数据库性能的"瓶颈"。

（4）检查操作系统资源，包括CPU、内存与I/O以及网络资源。

15.2 系统设计优化

一般来说，绝大多数系统效率的问题都是由设计上的缺陷造成的。如果Oracle设计合理，SQL程序效率较好，硬件可以满足用户的需求，DBA定期检视数据库的运行并及时地消除了影响系统效率的"瓶颈"，那么这个数据库系统就不会遇到效率的问题。

要理解系统的设计，首先必须了解数据库系统的分类。根据数据库系统的应用领域的不同可以将其分为两大类：联机事务处理（Online Transaction Processing，OLTP）系统和数据仓库（Data Warehouse，DW）系统/决策支持（Decision Support，DS）系统。表15-1所示为这两种系统的数据以及系统特性的比较。

表15-1 OLTP与DW/DS系统的数据以及系统特性的比较

联机事务处理系统	数据仓库/决策支持系统
面向应用	面向主题
详细	综合或提炼详细的数据
存取瞬间准确的数据	代表过去的数据
为日常工作服务	为管理者服务
可更新	不更新
重复运行	启发式运行
处理需求事先可知	处理需求事先不知道
对性能要求高	对性能要求低
一个时刻存取一个单元	一个时刻存取一个集合
事务处理驱动	分析处理驱动
更新控制主要涉及所有权	无更新控制问题
高可用性	松弛的可用性
整体管理	以子集管理
非冗余性	时常有冗余
静态结构：可变的内容	结构灵活
一次性处理数据量小	一次性处理数据量大
支持日常操作	支持管理需求
访问的高可用性	访问的低可能性或适度可能性

在介绍完毕联机事务处理和数据仓库的数据以及系统特性之后，下面简单地介绍一下这两种系统在设计上的不同要求。表15-2所示的是这两种系统的系统设计特点比较。

表15-2 OLTP与DW/DS系统的系统设计特点比较

序号	联机事务处理系统	数据仓库/决策支持系统
1	执行索引搜索	更多地执行全表扫描
2	使用B树索引	使用位图树索引

续表

序号	联机事务处理系统	数据仓库/决策支持系统
3	cursor_sharing 设为 SIMILAR	cursor_sharing 保留默认值 EXACT
4	一般不使用并行查询	对大的操作使用并行查询
5	pctfree 根据修改操作来设定	pctfree 可以设为 0
6	共享代码、使用绑定变量	使用常量和提示等

（1）联机事务处理系统的查询多数是查找满足某一或某些特定条件的数据，而这些数据只不过是整个表中很小的一个子集；而数据仓库上的查询多数是查找综合的数据（分类的数据），这样的数一般需要扫描整个表，这是因为管理者对个体的数据是不感兴趣的，他们只需要通过综合的信息来进行决策。

（2）由于联机事物处理系统的 DML 操作相当复杂，因此无法使用位图索引（位图索引的 DML 维护成本相当高）。查询这样的数据一般需要扫描整个表，因为管理者对个体的数据不感兴趣，他们只需要通过综合的信息来进行决策。

（3）在联机事物处理体系上将 cursor_sharing 设为 SIMILAR 的目的是最大限度地共享 SQL 代码。而在数据仓库系统上，代码编译在整个 SQL 语句执行过程中所占的时间比例非常小，而生成一个好的执行计划才是数据仓库系统最关心的问题，将 cursor_sharing 保留默认值 EXACT 可以帮助生成好的执行计划。

（4）联机事物处理系统上的查询数据量一般比校小；而数据仓库系统上的查询数据量一般都比较大。

（5）由于联机事物处理系统线上修改操作相当频繁，如果将 pctfree 设为 0，则会造成数据的迁移，使系统效率下降。数据仓库系统上的数据查询量一般都是一次性装入的，一般没有修改操作。

下面以参数 cursor_sharing 为例，介绍它的查看与设置操作。

步骤 1　查询参数 cursor_sharing，其代码如下：

```
SQL> show parameter cursor_sharing;

NAME                                 TYPE        VALUE
------------------------------------ ----------- --------------
cursor_sharing                       string      EXACT
```

步骤 2　修改参数 cursor_sharing，其代码如下：

```
SQL> alter system set cursor_sharing = similar;

系统已更改.
```

步骤 3　查询参数 cursor_sharing，其代码如下：

```
SQL> show parameter cursor_sharing;

NAME                                 TYPE        VALUE
------------------------------------ ----------- ---------------
cursor_sharing                       string      SIMILAR
```

15.3 内存优化

通过合理地分配内存大小,合理地设置表空间体系和内部空间参数,可以提高磁盘空间的利用率,减少数据段碎片,并且在查询和向数据文件写入数据时使用较少的 I/O。较少的 I/O 会降低 CPU 的资源消耗。同时,合理配置环境参数,可以使数据库中的数据顺畅地流动,以减少锁冲突和各种等待时间,充分地利用系统资源。

Oracle 数据库的内存组成为 SGA 与 PGA。当内存足够大时,内存配置不是大问题;但当主机的总内存不多或版本存在差异或碰到 SGA 有最大限额如 1.7GB 或 session 总数超过 2000 时,内存管理就是一个大问题了。因此,建议采用 SGA 与 PGA 分开优化的半自动内存管理策略。

15.3.1 SGA 内存优化

Oracle 的 SGA 是指系统全局区,它是数据库运行期间使用的一段公有内存,即所有使用数据库的用户都可以访问这部分内存。SGA 中最重要的 3 个组成部分如下所述:

(1) Database Buffer Cache:存储从实体文件中读出来的数据。

(2) Redo Log Buffer:存储用户改变数据的信息。当用户进行 Rollback 时就是靠 Redo Log Buffer 的数据进行回滚操作。

(3) Shared Pool:包含 Library Cache(存储执行 SQL 过程所需收集、分解、解析 SQL 的所有信息)、Dictionary Cache(存储有关于表、视图等结构的相关信息)、Result Cache(存储执行结果)。

1. SGA 信息

```
SQL> show sga

Total System Global Area 2550136752 bytes
Fixed Size                  9031600 bytes
Variable Size            1191182336 bytes
Database Buffers         1342177280 bytes
Redo Buffers                7745536 bytes
```

其中,第一个参数 Total System Global Area 的值为 2550136752 B(2432MB),其实就是 SGA 的最大值,它与 sga_max_size 大小相等,其代码如下:

```
SQL> show parameter sga_max_size;

NAME                                 TYPE         VALUE
------------------------------------ ------------ --------------
sga_max_size                         big integer  2432M
```

数据库高速缓冲区缓存(Database Buffer Cache)大小由参数 db_cache_size 指定。而重做日志缓冲区缓存(Redo Log Buffer Cache)大小通常由参数 log_buffer 来设置。下面以 log_buffer 为例进行细节说明。

步骤1　查看重做日志缓冲区大小，其代码如下：

```
SQL> show parameter log_buffer

NAME                                 TYPE         VALUE
------------------------------------ ------------ --------------
log_buffer                           big integer  7232k
```

该参数值的单位是 KB，一些版本的默认配置只有 512KB，这么小的重做日志缓冲区对繁忙的事务处理数据库系统来说显然是不够的。

步骤2　修改重做日志缓冲区大小，其代码如下：

```
SQL> alter system set log_buffer = 10240k scope = spfile;

系统已更改.
```

但是，如果立即使用查看重做日志缓冲区大小命令显示其大小，就会发现显示的值并没有变化。因为 log_buffer 这个参数是静态参数，若要修改它，就要对二进制参数文件 Spfile 文件进行修改，变化值要想发挥作用必须重新启动 Oracle 实例。

2. SGA 相关的静态参数

为了获取与 SGA 相关参数的当前值，首先要以用户 sys 登录数据库系统，并使用如下命令显示 SGA 相关参数信息。

```
SQL> show parameter sga

NAME                                 TYPE         VALUE
------------------------------------ ------------ --------------------
allow_group_access_to_sga            boolean      FALSE
lock_sga                             boolean      FALSE
pre_page_sga                         boolean      TRUE
sga_max_size                         big integer  2432M
sga_min_size                         big integer  0
```

```
sga_target                           big        integer     2432M
unified_audit_sga_queue_size                    integer     1048576
```

（1）sga_max_size：分配给 SGA 的最大内存。有时必须重新设置 sga_max_size 参数，如安装数据库使用的是默认设置，这个参数的值偏小。另外，如果在安装数据库之后，系统的物理内存增加了，SGA 就无法使用新增加的内存。这时，可以调整 SGA 的最大尺寸，目的是增加 Oracle 在整个内存中所占的比例，但是 SGA 的最大尺寸不能太大，一般可以设置为当前内存大小即可。

修改 sga_max_size 参数的命令如下：

```
SQL> alter system set sga_max_size = 3000m;
alter system set sga_max_size = 3000m
                 *
第 1 行出现错误：
ORA - 02095：无法修改指定的初始化参数
```

由于该参数是静态的，无法动态修改 sga_max_size 参数的值，即只能修改该参数在 spfile 中的值，修改后还要关闭数据库，重新启动后才有效。正确地修改该参数的命令如下：

```
SQL> alter system set sga_max_size = 3000m scope = spfile;
```

除了该参数之外，还有两个必须静态修改的与 SGA 有关的系统参数：lock_sga 和 pre_page_sga。

（2）lock_sga：lock_sga 参数的作用是将 SGA 锁定在物理内存内，这样就不会发生 SGA 使用虚拟内存的情况，显然这样可以提高数据的读取速度。注意，磁盘 I/O 操作应尽量避免或者减少。该参数的默认值是 FALSE，即不将 SGA 锁定在内存中。由于 lock_sga 参数为静态参数，所以无法动态地修改 lock_sga 参数的值。修改 lock_sga 参数正确的命令如下：

```
SQL> alter system set lock_sga = true scope = spfile;
```

（3）pre_page_sga：该参数的作用是在启动数据库实例时，将整个 SGA 读入物理内存。这样做虽然增加了实例启动的时间和所需的物理内存，但是对于内存充足的系统而言，这样显然可以提高系统的运行效率。该参数的默认值是 FALSE，若要修改参数值为 TRUE，正确的命令如下：

```
SQL> alter system set pre_page_sga = true scope = spfile;
```

（4）sga_target：上述参数为静态参数，所以不能动态调整，而其他多数参数都可以在不关闭实例的情况下动态地调整。不过这对 Oracle 使用人员来说仍然不是一件容易的事。那么，能不能让 Oracle 系统根据系统的实际需要自动调整 SGA 各个部分的大小呢？答案是肯定的。这个参数就是 sga_target，它就是主管 SGA 自动管理的。若该参数为 0，则代表没有启动 SGA 自动管理，可以通过以下命令修改该参数的值。

```
SQL> alter system set sga_target = 512m;
```

参数 sga_target 的值是可以动态修改的,而且可以设置到与 sga_max_size 同样大小。

3. SGA 可以自动调整的内存组件

虽然 SGA 可以自动管理,但不是所有的内存组件都可以自动调整。那么,哪些 SGA 是可以自动调整的呢? 下面罗列 5 个可以自动调整的 SGA 内存组件。

(1) 共享池。
(2) Java 池。
(3) 大池。
(4) 数据库缓冲区。
(5) 流池。

这些组件的尺寸不需要用户干预,其值自动设置为 0,使用视图 v$parameter 便可以查看这些自动调整的内存组件的信息,其代码如下:

```
SQL> select name,value ,isdefault
  2  from v$parameter
  3  where name in('shared_pool_size','large_pool_size','java_pool_size');

NAME                              VALUE                ISDEFAULT
-------------------------------- -------------------- --------------
shared_pool_size                  0                    TRUE
large_pool_size                   0                    TRUE
java_pool_size                    0                    FALSE
```

虽然这些参数可以自行调整,但用户依然可以使用 alter system set 语句修改组件的尺寸。例如下面的代码可以将 Java 池大小修改为 10MB。

```
SQL> alter system set java_pool_size = 10m;

系统已更改.
```

使用下列指令可以查询参数 java_pool_size 的修改结果。

```
SQL> show parameter java_pool_size;

NAME                                 TYPE         VALUE
------------------------------------ ------------ --------------
java_pool_size                       big integer  16MB
```

认真观察上述结果,发现参数的值并没有改变为 10MB,这是因为 Oracle 会自动对系统自身情况做一些调整。

15.3.2 PGA 内存优化

PGA 是程序全局区。在数据库会话专有连接模式下,该区域是私有区域,服务器进程与用户进程一一对应。在共享服务器会话连接模式下,一个服务器进程对应各用户进程。PGA 是多个用户进程共享使用。

所谓的 PGA 优化就是将这些大规模的数据排序放在 PGA 中运行,即 Oracle 为了加快排序的速度,在内存中设有排序区,而不是使用虚拟内存而占用操作系统的交换区。

1. PGA 排序区自动管理

目前的版本都支持内存排序区的自动化管理。要使用内存排序区的自动管理这一功能,必须使用两个 Oracle 系统参数:pga_aggregate_target 和 workarea_size_policy。

使用如下代码可查看 PGA 排序区是否为自动管理。

```
SQL> select name,value,isdefault
  2  from v$parameter
  3  where name in('pga_aggregate_target','workarea_size_policy');

NAME                           VALUE                      ISDEFAULT
------------------------------ -------------------------- -----------
pga_aggregate_target           847249408                  FALSE
workarea_size_policy           AUTO                       TRUE
```

(1) pga_aggregate_target:参数 pga_aggregate_target 定义了实例中所有服务器进程的 PGA 内存的总和。它的定义范围是 10MB~4000GB。从上述代码中发现,其 ISDEFAULT 的值为 FALSE,说明它的值是需要设置的,它不是系统的确认参数。使用 alter system set 的命令修改 pga_aggregate_target 的值,但是修改为多大值呢?这需要分析一下当前数据库的 PGA 状态。

(2) workarea_size_policy:workarea_size_policy 的值可以为 AUTO 或者 MANUAL。上述代码中该参数的值为 AUTO,代表排序方式为自动管理,否则为手动完成。

2. PGA 状态

通过动态性能视图 v$pgastat 查询 PGA 状态信息,其代码如下:

```
SQL> select * from v$pgastat;

NAME                                     VALUE        UNIT        CON_ID
---------------------------------------- ------------ ----------- ----------
aggregate PGA target parameter           847249408    bytes       0
aggregate PGA auto target                543845376    bytes       0
global memory bound                      104857600    bytes       0
```

```
total PGA inuse                              242974720    bytes      0
total PGA allocated                          316269568    bytes      0
maximum PGA allocated                        362841088    bytes      0
total freeable PGA memory                     33423360    bytes      0
MGA allocated (under PGA)                            0    bytes      0
maximum MGA allocated                                0    bytes      0
process count                                       80               0
max processes count                                 82               0
PGA memory freed back to OS                   65011712    bytes      0
total PGA used for auto workareas                    0    bytes      0
maximum PGA used for auto workareas            2243584    bytes      0
total PGA used for manual workareas                  0    bytes      0
maximum PGA used for manual workareas                0    bytes      0
over allocation count                                0               0
bytes processed                              167824384    bytes      0
extra bytes read/written                             0    bytes      0
cache hit percentage                               100    percent    0
recompute count (total)                           1001               0
```

已选择 21 行.

以上输出中记录 PGA 状态的 3 个参数，其含义如下所示。

（1）aggregate PGA target parameter：用户设置的当前 PGA 的内存总和，该系数将 byte 换算成为 MB，为 808MB。

（2）aggregate PGA auto target：Oracle 为 PGA 的排序区分配的内存大小，显然其值不能超过 PGA 内存的总和，上述代码中为 516.4541MB。

（3）cache hit percentage：排序区完成的比例，100 percent 表示全部排序都在 PGA 完成，所以 Oracle 自动分配的排序区的尺寸是合理的。如果出现部分排序不在 PGA 的排序区完成，则 value 的值将小于 100，需要考虑适当增加 PGA 的内存总和。

3. 调整 PG

调整 PGA 内存大小可以通过下列语句执行。

```
SQL> alter system set  PGA_aggregate_target = 1000m;
```

系统已更改.

修改后可以通过 pga_aggregate_target 参数查看修改后的结果。

```
SQL> show parameter pga_aggregate_target;

NAME                                 TYPE         VALUE
------------------------------------ ------------ --------------
pga_aggregate_target                 big integer  1000M
```

通过性能视图 v$pgastat 可以查看 PGA 的大小以及 PGA 中排序区的大小。

```
SQL> select * from v$pgastat where name in('aggregate PGA target parameter','aggregate PGA auto
target');

NAME                                    VALUE       UNIT                CON_ID
------------------------------------    ----------  ----------------    ----------
aggregate PGA target parameter          1048576000  bytes                        0
aggregate PGA auto target                724893696  bytes                        0
```

结果发现，PGA 的总和已经修改为 104857600byte(1000MB)，同时自动分配给 PGA 的排序区的尺寸为 724893696byte(688.9394MB)。由此可见，随着 PGA 的总内存的增加，Oracle 会自动增加其为排序区分配的内存容量。

15.4 I/O 优化

数据库的作用就是实现对数据的管理和查询。任何一个数据库系统，必然存在对数据的大量读和写的操作，即 I/O 操作。I/O 问题也往往是导致数据库性能问题的重要原因。一般数据库进行了内存优化之后，多数情况下数据库的性能是可以达到用户的要求的。但是有时经过这些优化过程之后，系统性能仍然没有明显的改善，这时候就需要进行 I/O 优化操作了。

15.4.1 Oracle 中 I/O 的产生

I/O 当然包括了读、写两部分，先介绍 Oracle 中读与写操作的产生。

先简单地看下 Oracle 的物理结构。Oracle 的物理文件包括以下 3 种文件：控制文件、重做日志文件、数据文件。而根据功能的不同，数据文件还可以分为系统数据文件、临时空间文件、回滚段文件和用户数据文件。另外，如果数据库的归档模式被激活，还存在归档日志文件。Oracle 的 I/O 产生，就是对这些文件的数据读、写操作。

为了使读者更好地理解 Oracle 数据库的 I/O 操作，从表 15-3 给出主要后台进程对不同种类文件的 I/O 操作。

表 15-3 后台进程的 I/O 操作

进程	Oracle 文件的 I/O			
	数据文件	重做日志文件	归档日志文件	控制文件
CKPT				读/写
DBWn	写			
LGWR		写		读/写
ARCn		读	写	读/写
SERVER	读/写	读		读/写

各种文件的增长速度比较如表 15-4 所示。

表 15-4 各种文件增长速度比较

Disk space consumption area	Expected growth
Oracle Archived Redo Log Files	High
Oracle Data Files	High
Oracle Index Files	High
Oracle Control Files	Mediun
Oracle Initialization Files	Small

由 Oracle 数据库物理存储架构、数据库物理文件增长情况和测试发现,给 I/O 带来很大负荷的是重做日志文件、数据文件与索引文件的读写。当数据量、性能和持续压力时间都达到一定量时,这三者产生的 I/O 负荷相当,因此降低磁盘 I/O 的最终解读就是降低这 3 类文件的 I/O。

15.4.2 I/O 竞争文件与数据的优化

1. I/O 文件的分布原则和方法

(1) 数据文件与重做日志文件最好分别放在不同的硬盘上,特别是并行操作多的联机事务处理系统。从表 15-3 可以看出,有 3 个进程要操作数据文件,而且数据文件的 I/O 量也是最大的;同样也有 3 个进程要操作重做日志文件,因此数据文件与重做日志文件的 I/O 竞争非常大。

(2) 归档日志文件与重做日志文件放在不同的硬盘上。

(3) 不同的重做日志(成员)文件放在不同的硬盘上。

(4) 不同的控制文件最好也分别放在不同的硬盘上。

(5) 不同的归档日志文件最好也放在不同的硬盘上。

(6) 不同表空间对应的文件分别放在不同的硬盘上。

(7) 不同类型的数据存放在不同的表空间中。

(8) 减少与数据库无关操作的磁盘 I/O。

2. I/O 竞争数据的分布原则和方法

(1) 表空间 SYSTEM 应该只存放用户 sys 和数据字典对象,其他任何用户的对象都不应该放入该表空间。

(2) 还原表空间只能由还原段使用,其他不能存放任何其他对象。

(3) 临时表空间很容易被碎片化,所以临时表空间只能作为排序区使用,其中也不能存放其他对象。

(4) 表和索引分别存放在不同的表空间中,因为一个表和表上的索引是明显存在竞争的。

(5) 大对象(LOB)应该放在单独的表空间 LOB。

15.4.3 I/O 物理文件的优化

1. 文件监控与诊断

首先了解一个动态性能视图 v$filestat。利用这个视图可以找到每个数据文件的 I/O 量。

```
SQL> desc v$filestat;
名称                                      是否为空? 类型
----------------------------------------- -------- ----------------
FILE#                                              NUMBER
PHYRDS                                             NUMBER
PHYWRTS                                            NUMBER
PHYBLKRD                                           NUMBER
OPTIMIZED_PHYBLKRD                                 NUMBER
PHYBLKWRT                                          NUMBER
SINGLEBLKRDS                                       NUMBER
READTIM                                            NUMBER
WRITETIM                                           NUMBER
SINGLEBLKRDTIM                                     NUMBER
AVGIOTIM                                           NUMBER
LSTIOTIM                                           NUMBER
MINIOTIM                                           NUMBER
MAXIORTM                                           NUMBER
MAXIOWTM                                           NUMBER
CON_ID                                             NUMBER
```

其中,file#表示数据文件号;phyblkrd 表示所读的物理数据库块的块数;phyblkwrt 表示所写的数据物理块的块数;readtim 表示读操作所使用的时间(ms);writetim 表示写操作所使用的时间(ms)。

此外,需查看参数 timed_statistics 的值是否为 TRUE,因为如果 timed_statistics 的值是 FALSE,那么 readtim 和 writetim 将都为 0,其代码如下:

```
SQL> show parameter timed_statistics;

NAME                                 TYPE        VALUE
------------------------------------ ----------- ------------
timed_statistics                     boolean     TRUE
```

然后就可以通过查询该动态性能视图查看当前文件的 I/O 操作,其代码如下:

```
SQL> select file#,phyblkrd,phyblkwrt,readtim,writetim
  2  from v$filestat;
```

```
    FILE#   PHYBLKRD   PHYBLKWRT   READTIM   WRITETIM
--------- ---------- ----------- --------- ----------
        1       8393         390       236          3
        2         13           6         0          0
        3       5573        1710       162         65
        4        589         810        15         38
        7         13           6         0          0
```

利用 Oracle 的数据字典 v$filestat 可以追踪数据文件以及 I/O 操作。如果 I/O 的问题是由多个 I/O 量大的数据文件存放在一块硬盘引起的,就可以用之前介绍的办法将数据文件分别移动到不同的物理硬盘上。但是有时候 I/O 的问题也可能是表和索引以及 I/O 的问题。

2. 表的 I/O 优化

表的 I/O 问题有可能是数据行迁移的问题,这个之前介绍过,此处就不再多举例说明了。简单介绍一下步骤。

步骤 1 查询表的数据行迁移状况,需要先做统计。

(1) analyze table xxx. yyy compute statistics;

(2) select num_rows, chain_cnt from dba_tables where owner = 'XXX' and table_name = 'YYY';

注意''中的内容需要大写。其中 chain_cnt 字段就是迁移数量。

步骤 2 如果迁移比较严重,避免后面继续发生,所以要加大 pctfree 的参数。

(1) 首先确认当前值。

```
select table_name, pct_free from dba_tables where owner = 'XXX' and table_name = 'YYY';
```

(2) 如需要的话可以进行调整。

```
alter table xxx.yyy pctfree 30;
```

3. 索引 I/O 优化

当某个大型表或索引引起 I/O"瓶颈",那就涉及了索引优化问题。索引可能存在左右不平衡或大量被标记删除却并未真正删除的索引,从而影响性能。

步骤 1 收集信息,其代码如下:

```
SQL> execute dbms_stats.gather_index_stats('用户名','索引名');
```

或

```
SQL> analyze index xxx.iii compute statistics;
```

步骤 2 查询索引删除相关信息,其代码如下:

```
SQL> select name, (del_lf_rows_len/lf_rows_len) * 100 as wastage from index_stats;
```

其中:
lf_rows 表示当前索引记录行数;
lf_rows_len 表示全部索引长度;
del_lf_rows 表示删除的索引记录行数;
del_lf_rows_len 表示被删除索引的全部长度。
步骤3　查询索引深度信息,其代码如下:

```
SQL> select index_name, num_rows, blevel, status from dba_indexes where table_owner = 'XXX' and
table_name = 'YYY';
```

一般地,blevel 的值在 3 以下都没问题。如果大于 3,就需要重建索引了。
步骤4　重建索引消除问题。有时需要加大数据块(db_block_size)大小来降低索引深度,推荐 8KB。

```
SQL> alter index xxx.iii rebuild;
```

4. 重做日志文件的优化

之前介绍的数据库问题和解决方法都集中在数据文件上,那么重做日志文件的配置会不会造成数据库系统的效率问题呢?答案是当然会。如果重做日志文件设置得太小,就会造成重做日志组的频繁切换,而且重做日志组在切换时还要产生检查点。在检查点时除了要进行读写数据文件的文件头和控制文件之外,还要调用 DWRn 后台进程,将所有的脏数据写到数据文件中。这就会产生大量的输入和输出,所以重做日志文件不能设置得太小。

接下来给大家演示怎样获得重做日志切换的时间以及时间间隔。使用数据字典 v$log_history 获取重做日志切换的时间及时间间隔的命令如下:

```
SQL> select sequence#, to_char(first_time, 'RR-MM-DD HH:MM:SS') "Date Time" from v$log_
history;

SEQUENCE#   Date Time
---------   -----------------------------
        1   20-02-19 06:02:47
        1   20-02-19 07:02:00
        2   20-02-20 05:02:13
        3   20-02-20 05:02:34
        1   20-02-20 05:02:08
        2   20-02-20 06:02:21
        3   20-02-20 06:02:37
        1   20-02-20 06:02:44
        2   20-02-21 11:02:00
        1   20-02-23 01:02:32

已选择 10 行。
```

15.4.4 I/O 调优的其他手段

通过对 AWR 报告的分析,可以得知到底是哪些 I/O 相关事件引起的 I/O 问题。针对不同的事件,可以采取不同的分析与处理方法。而有一些通用的方法并不是针对特定的事件的。这里先介绍一下这些通用方法。

1. 通过 SQL 调优来减少 I/O 请求

一个没有任何用户 SQL 的数据库几乎不产生任何 I/O。基本上数据库所有的 I/O 都是直接或间接由用户提交的 SQL 所导致的。这意味着可以通过控制单个 SQL 产生的 I/O 来降低数据库总的 I/O 请求。而通过 SQL 调优来降低 SQL 查询计划中的 I/O 操作次数则是降低 SQL 产生 I/O 的最好方法。数据库的性能问题通常是由少数几个 SQL 语句所导致的,它们产生了大量 I/O 导致了整个数据库的性能下降。优化几条问题语句往往就能解决整个数据库的 I/O 性能问题。

从 Oracle 10g 版本开始,ADDM 能够自动检测出问题语句,同时,再通过查询优化建议器能够自动优化语句并降低它们对 I/O 的消耗。

2. 通过调整实例参数来减少 I/O 请求

在这种方法中,主要有两种途径来实现对 I/O 的优化。

(1) 使用内存缓存来减少 I/O。通过一些内存缓存,如 Buffer Cache、Log Buffer、Sort Area,可以降低数据库对 I/O 的请求。当 Buffer Cache 增加到一定大小时,绝大多数的结果可以直接从缓存中获取,而无须从磁盘上读取。而在进行排序操作时,如果 Sort Area 足够大,排序过程中产生的临时数据可以直接放在内存中,而无须占用临时表空间。

(2) 调整多数据块 I/O 的大小。控制多数据块 I/O 的参数称作 db_file_multiblock_read_count,它能够控制在多数据块读时一次读入数据块的次数。适当增加这个参数大小,可以提高多数据块操作(如全表扫描)的 I/O 效率。例如,读取 100MB 数据,如果每次读取 1MB 的数据 100 次的效率就比每次读取 100KB 数据 1000 次更快。但是这个数字达到一定大小后,再增加就作用不明显了:每次 10MB 一共读 100 次来读取 1GB 的数据的效率和单独一次读取 1GB 数据的效率是没有多大区别的。这是因为 I/O 效率受到两个因素的影响: I/O 建立时间和 I/O 传输时间。

① I/O 建立时间对于不同 I/O 大小来说都是相同的,它决定了对小 I/O 的总的 I/O 时间,增大多数据块 I/O 大小可以减少 I/O 建立时间。

② I/O 传输时间与 I/O 大小是成正比的,在小 I/O 时,I/O 传输时间一般比 I/O 建立时间少,但对于大 I/O 操作来说,I/O 传输时间决定了总的 I/O 时间。因此,多数据块 I/O 大小增大到一定大小时,它对总的 I/O 时间影响就不大了。

3. 在操作系统层面优化 I/O

如前面所介绍的,利用一些操作系统提供的提升 I/O 性能的特性,如文件系统的异步 I/O、Direct I/O 等来优化数据库系统的 I/O 性能。另外一种方法就是增加每次传输的最大 I/O 大小的限制(大多数的 UNIX 系统中,由参数 max_I/O_size 控制)。

4. 通过 Oracle ASM 实现对 I/O 的负载均衡

自动存储管理(Automatic Storage Manager,ASM)是从 Oracle 10g 版本开始引入的。

它是一个建立在数据库内核中的文件系统和卷管理器。它能自动将 I/O 负载均衡到所有可用的磁盘启动器上去,以避免"热区"。ASM 能防止碎片,因此无须重建数据就可以回收空间。数据能被均衡分布到所有硬盘上。

5. 通过条带化、RAID、SAN 或者 NAS 实现对 I/O 的负载均衡

这个方法通过一些成熟的存储技术(如条带化、RAID、SAN 和 NAS),将数据库 I/O 分布到多个可用的物理磁盘,从而实现负载均衡,以避免存在空闲可用磁盘情况下仍出现磁盘争用的情况和 I/O"瓶颈"问题。

6. 通过手工布置数据库文件到不同的文件系统、控制器和物理设备上来重新分布数据库 I/O

当数据库系统中缺乏以上各种存储技术手段时,可以考虑使用这种方式。这样做的目的是使数据库的 I/O 得到均衡分布,从而避免在还有空闲磁盘时出现磁盘争用和 I/O"瓶颈"问题。当然这种手动分布 I/O 方法是无法达到以上的自动分布 I/O 的效果的。

7. 其他手段

系统中总会存在一些 I/O 是无法消除或降低的。如果采用以上手段还不能满足 I/O 性能要求的话,可以考虑以下两种方法。

(1) 将旧数据移除你的生产数据库。
(2) 采用更多、更快的硬件。

15.5 SQL 语句优化

15.5.1 SQL 语句优化原则

随着数据库中数据的增加,系统的响应速度就成为目前系统需要解决的主要问题之一,系统优化中一个很重要的方面就是 SQL 语句的优化。大量的数据指出劣质的 SQL 语句和优质的 SQL 语句之间的速度差别可以达到上百倍,对一个系统不是简单的能实现功能就可以,而是要写出高质量的 SQL 语句提高系统的可用性。

总的来说,SQL 语句的编写原则和 SQL 语句的优化主要包含以下 4 个方面。

(1) 不要让 Oracle 进行太多的操作。
① 避免复杂的多表连接。
② 避免使用"*"列出所有列。
③ 避免使用消耗资源的操作,如 distinct、union、minus、intersect、order by。
④ 用 exitsts 替换 distinct。
⑤ 用 union_all 替换 union。

(2) 给优化器更明确的命令。
① 自动选择索引。
② 至少包含组合索引的第一列。
③ 避免在索引列上使用函数。
④ 避免使用通配符。
⑤ 避免在索引列上使用 not。

⑥ 避免在索引列上使用 is null 和 is not null。
⑦ 避免出现索引列自动转换。
⑧ 在查询时尽量少用格式转换。
(3) 减少访问次数。
① 使用 decode 来减少处理时间。
② 在含有子查询的 SQL 语句中,要特别注意减少对表的查询。
(4) 注意细节上的影响。
① Oracle 采用自下而上的顺序解析 where 子句,根据这个原理,注意 where 子句中的连接顺序。即 where 子句中排在最后的表应当是返回行数最少的表,有过滤条件的子句应该放在 where 子句的最后。
② 最好不要在 where 子句中使用函数或者表达式。
③ 用 where 子句替换 having 子句。
④ 用 not exists 替换 not in。
⑤ 用索引提高效率。
⑥ 避免在索引上使用计算。
⑦ 用 >= 代替 >。
⑧ 通过使用 >= 和 <= 等,避免使用 not 命令。
⑨ 利用外部连接+代替效率低下的 not in 运算。
⑩ 尽量多使用 commit。
⑪ 用 truncate 代替 delete。
⑫ 注意字符型字段的引号。

15.5.2　SQL 语句执行计划

如何发现最消耗系统资源的 SQL 语句呢？利用 I/O 优化鉴定耗费资源的方法即 AWR 报告,即可使之变得很容易。当确定了一个最有可能出问题的 SQL 语句后,接下来就是如何追踪这一语句的执行,最普遍的方法就是找出该 SQL 语句的执行计划。

那么执行计划又是什么呢？简言之,一个执行计划就是在执行一条 SQL 语句和相关的操作时,优化器所执行的操作步骤。一个执行计划包括了该语句须访问的每一个表的访问方法以及访问这些表的顺序。观察执行计划本身已经清楚地解释：为什么这个语句效率如此之低。下面简单介绍几种常用而简单易学的获取 SQL 语句执行计划的方法。

1. 利用 autotrace 追踪 SQL 语句

利用 autotrace 追踪 SQL 语句的代码如下：

```
SQL> set autotrace on
SQL> select * from dual;


D
-
X

执行计划
----------------------------------------------
```

```
Plan hash value: 272002086

--------------------------------------------------------------------------
| Id  | Operation         | Name | Rows  | Bytes | Cost (%CPU)| Time     |
--------------------------------------------------------------------------
|   0 | SELECT STATEMENT  |      |     1 |     2 |     2   (0)| 00:00:01 |
|   1 |  TABLE ACCESS FULL| DUAL |     1 |     2 |     2   (0)| 00:00:01 |
--------------------------------------------------------------------------
```

统计信息
```
----------------------------------------------------------
          1  recursive calls
          0  db block gets
          2  consistent gets
          0  physical reads
          0  redo size
        547  bytes sent via SQL*Net to client
        381  bytes received via SQL*Net from client
          2  SQL*Net roundtrips to/from client
          0  sorts (memory)
          0  sorts (disk)
          1  rows processed
```

执行完语句后，会显示执行计划与统计信息。在用该方法查看执行时间较长的 SQL 语句时，需要等待该语句执行成功后，才能返回执行计划，使优化的周期大大的增加了。如果不想执行语句，而只是想得到执行计划，可以采用如下命令。

```
SQL> select * from dual;
```

执行计划
```
----------------------------------------------------------
Plan hash value: 272002086

--------------------------------------------------------------------------
| Id  | Operation         | Name | Rows  | Bytes | Cost (%CPU)| Time     |
--------------------------------------------------------------------------
|   0 | SELECT STATEMENT  |      |     1 |     2 |     2   (0)| 00:00:01 |
|   1 |  TABLE ACCESS FULL| DUAL |     1 |     2 |     2   (0)| 00:00:01 |
--------------------------------------------------------------------------
```

统计信息
```
----------------------------------------------------------
          0  recursive calls
          0  db block gets
          2  consistent gets
          0  physical reads
          0  redo size
        547  bytes sent via SQL*Net to client
```

```
      381  bytes received via SQL*Net from client
        2  SQL*Net roundtrips to/from client
        0  sorts (memory)
        0  sorts (disk)
        1  rows processed
```

上述的代码只列出了执行计划,而不会真正地执行语句,这就大大优化了时间。虽然也列出了统计信息,但是因为没有执行语句,所以该统计信息是没有用处的。

也可以使用如下指令单纯地显示执行计划。

```
SQL> set autotrace traceonly explain
SQL> select * from dual;

执行计划
----------------------------------------------------------
Plan hash value: 272002086

--------------------------------------------------------------------------
| Id  | Operation         | Name | Rows  | Bytes | Cost (%CPU)| Time     |
--------------------------------------------------------------------------
|   0 | SELECT STATEMENT  |      |     1 |     2 |     2   (0)| 00:00:01 |
|   1 |  TABLE ACCESS FULL| DUAL |     1 |     2 |     2   (0)| 00:00:01 |
--------------------------------------------------------------------------
```

当然也可以只显示统计信息,代码如下:

```
SQL> set autotrace traceonly statistics;
SQL> select * from dual;

统计信息
----------------------------------------------------------
        0  recursive calls
        0  db block gets
        2  consistent gets
        0  physical reads
        0  redo size
      547  bytes sent via SQL*Net to client
      380  bytes received via SQL*Net from client
        2  SQL*Net roundtrips to/from client
        0  sorts (memory)
        0  sorts (disk)
        1  rows processed
```

追踪完毕 SQL 语句之后,可以使用如下命令关闭自动追踪功能。

```
SQL> set autotrace off
SQL> show autotrace
autotrace OFF
```

2. 利用 explain plan 命令

explain plan 命令被用来产生一个优化器所使用的执行计划，这一命令将所产生的执行计划存储在一个表中，系统默认为表 plan 中，但是该命令并不真正地执行语句，而只是产生可能使用的执行计划，如果仔细地观察这一执行计划，就可以了解到服务器是如何执行所解释的 SQL 语句。

explain plan 命令的语法如下：

```
explain plan set statement_id = '正文'
into 用户名.表名
for 语句
```

其中，正文表示为语句的标示符是一个可选项。用户名.表名表示存放执行计划的表名，默认为 plan_table，这也是一个可选项。语句表示要解释的 SQL 语句正文。

15.6 项目案例

15.6.1 SQL 语句跟踪与优化实例

步骤 1 以用户名 sys 登录，查询用户 hr 下表 departments 与表 employees 连接的数据，其代码如下：

```
SQL> conn sys as sysdb
输入口令：
已连接.
SQL> select e.last_name,d.department_name
  2  from   hr.employees e,hr.departments d
  3  where e.department_id = d.department_id;

LAST_NAME                DEPARTMENT_NAME
------------------------ --------------------
Whalen                   Administration
Fay                      Marketing
Hartstein                Marketing
Tobias                   Purchasing
Colmenares               Purchasing
......

已选择 106 行.
```

步骤 2　开启追踪，其代码如下：

```
SQL> show autotrace
autotrace OFF
SQL> set autotrace on
```

步骤 3　显示语句的执行计划和统计信息，其代码如下：

```
SQL> select e.last_name,d.department_name
  2  from   hr.employees e, hr.departments d
  3  where e.department_id = d.department_id;

LAST_NAME                 DEPARTMENT_NAME
----------------------------------------------------
Whalen                    Administration
......
已选择 106 行.

执行计划
----------------------------------------------------
Plan hash value: 1473400139
```

Id	Operation	Name	Rows	Bytes	Cost (%CPU)	Time
0	SELECT STATEMENT		106	2862	5 (20)	00:00:01
1	MERGE JOIN		106	2862	5 (20)	00:00:01
2	TABLE ACCESS BY INDEX ROWID	DEPARTMENTS	27	432	2 (0)	00:00:01
3	INDEX FULL SCAN	DEPT_ID_PK	27		1 (0)	00:00:01
*4	SORT JOIN		107	1177	3 (34)	00:00:01
5	VIEW	index$_join$_001	107	1177	2 (0)	00:00:01
*6	HASH JOIN					
7	INDEX FAST FULL SCAN	EMP_DEPARTMENT_IX	107	1177	1 (0)	00:00:01
8	INDEX FAST FULL SCAN	EMP_NAME_IX	107	1177	1 (0)	00:00:01

```
Predicate Information (identified by operation id):
----------------------------------------------------
  4 - access("E"."DEPARTMENT_ID"="D"."DEPARTMENT_ID")
      filter("E"."DEPARTMENT_ID"="D"."DEPARTMENT_ID")
  6 - access(ROWID=ROWID)

统计信息
----------------------------------------------------
          1  recursive calls
          0  db block gets
         20  consistent gets
          0  physical reads
```

```
         0  redo size
      3340  bytes sent via SQL*Net to client
       556  bytes received via SQL*Net from client
         9  SQL*Net roundtrips to/from client
         1  sorts (memory)
         0  sorts (disk)
       106  rows processed
```

步骤4　关闭 trace 自动跟踪功能,其代码如下:

```
SQL> set autotrace off
SQL> show autotrace
autotrace OFF
```

步骤5　使用 explain plan 命令获取执行计划,默认存放在 plan_table 表中,其代码如下:

```
SQL> explain plan set statement_id = 'test_project'
  2  for
  3  select e.last_name, d.department_name
  4  from   hr.employees e, hr.departments d
  5  where e.department_id = d.department_id;

已解释.
```

步骤6　执行成功后,可以使用如下命令显示 plan_table 表中的所有执行计划了,其代码如下:

```
SQL> select id, operation, options, object_name, statement_id
  2  from plan_table;

ID    OPERATION       OPTIONS           OBJECT_NAME          STATEMENT_ID
----- --------------- ----------------- -------------------- ------------
  0   SELECT          STATEMENT                              test_project
  1   MERGE           JOIN                                   test_project
  2   TABLE           ACCESS            BY INDEX ROWID       DEPARTMENTS          test_project
  3   INDEX                             FULL SCAN            DEPT_ID_PK           test_project
  4   SORT            JOIN                                   test_project
  5   VIEW                                                   index$_join$_001     test_project
  6   HASH            JOIN                                   test_project
  7   INDEX                             FAST FULL SCAN       EMP_DEPARTMENT_IX    test_project
  8   INDEX                             FAST FULL SCAN       EMP_NAME_IX          test_project

已选择 9 行.
```

步骤 7 当以上的 explain plan 命令执行后,就可使用如下命令利用 dbms_xplan 软件包中 display 函数来显示 explain plan 命令的输出结果。

```
SQL> select plan_table_output from table(DBMS_XPLAN.DISPLAY());

PLAN_TABLE_OUTPUT
--------------------------------------------------------------------------------
Plan hash value: 1473400139
--------------------------------------------------------------------------------
| Id  | Operation                     | Name                | Rows | Bytes | Cost (%CPU) | Time     |
--------------------------------------------------------------------------------
|  0  | SELECT STATEMENT              |                     |  106 |  2862 |   5  (20)   | 00:00:01 |
|  1  |  MERGE JOIN                   |                     |  106 |  2862 |   5  (20)   | 00:00:01 |
|  2  |   TABLE ACCESS BY INDEX ROWID | DEPARTMENTS         |   27 |   432 |   2   (0)   | 00:00:01 |
|  3  |    INDEX FULL SCAN            | DEPT_ID_PK          |   27 |       |   1   (0)   | 00:00:01 |
|* 4  |   SORT JOIN                   |                     |  107 |  1177 |   3  (34)   | 00:00:01 |
|  5  |    VIEW                       | index$_join$_001    |  107 |  1177 |   2   (0)   | 00:00:01 |
|* 6  |     HASH JOIN                 |                     |      |       |             |          |
|  7  |      INDEX FAST FULL SCAN     | EMP_DEPARTMENT_IX   |  107 |  1177 |   1   (0)   | 00:00:01 |
|  8  |      INDEX FAST FULL SCAN     | EMP_NAME_IX         |  107 |  1177 |   1   (0)   | 00:00:01 |
--------------------------------------------------------------------------------

Predicate Information (identified by operation id):
---------------------------------------------------

   4 - access("E"."DEPARTMENT_ID" = "D"."DEPARTMENT_ID")
       filter("E"."DEPARTMENT_ID" = "D"."DEPARTMENT_ID")
   6 - access(ROWID = ROWID)

已选择 22 行。
```

步骤 8 为了方便 DBA 工作,Oracle 还提供了一个名为 utlxpls.sql 的脚本文件,它的功能就是显示最后解释的 SQL 语句的执行计划。这个脚本文件存放在 $ORACLE_HOME\RDBMS\ADMIN\ 文件夹中,可以使用如下命令执行这一个脚本,效果与之前完全一样。

```
SQL> @D:\software\oracle19c\rdbms\admin\utlxpls.sql

PLAN_TABLE_OUTPUT
--------------------------------------------------------------------------------
Plan hash value: 1473400139
--------------------------------------------------------------------------------
| Id  | Operation                     | Name                | Rows | Bytes | Cost (%CPU) | Time     |
--------------------------------------------------------------------------------
|  0  | SELECT STATEMENT              |                     |  106 |  2862 |   5  (20)   | 00:00:01 |
|  1  |  MERGE JOIN                   |                     |  106 |  2862 |   5  (20)   | 00:00:01 |
|  2  |   TABLE ACCESS BY INDEX ROWID | DEPARTMENTS         |   27 |   432 |   2   (0)   | 00:00:01 |
|  3  |    INDEX FULL SCAN            | DEPT_ID_PK          |   27 |       |   1   (0)   | 00:00:01 |
|* 4  |   SORT JOIN                   |                     |  107 |  1177 |   3  (34)   | 00:00:01 |
|  5  |    VIEW                       | index$_join$_001    |  107 |  1177 |   2   (0)   | 00:00:01 |
|* 6  |     HASH JOIN                 |                     |      |       |             |          |
|  7  |      INDEX FAST FULL SCAN     | EMP_DEPARTMENT_IX   |  107 |  1177 |   1   (0)   | 00:00:01 |
|  8  |      INDEX FAST FULL SCAN     | EMP_NAME_IX         |  107 |  1177 |   1   (0)   | 00:00:01 |
```

```
------------------------------------------------------------
Predicate Information (identified by operation id):
------------------------------------------------------------
   4 - access("E"."DEPARTMENT_ID"="D"."DEPARTMENT_ID")
       filter("E"."DEPARTMENT_ID"="D"."DEPARTMENT_ID")
   6 - access(ROWID=ROWID)

已选择 22 行。
```

其实打开这个文件会发现,这个脚本使用的方法也是调用了 dbms_xplan 软件包中的 display() 函数。在 utlxpls.sql 文件中调用 display() 函数时还用了几个参数,完整的 display() 函数调用格式如下:

```
select plan_table_output from
table(dbms_xplan.display('plan_table',NULL,'serial'));
```

其中:

plan_table: 存放 SQL 与执行计划的表;

serial: 不显示并行操作的信息。

步骤 9 利用数据字典 v$sql 获取了上面 SQL 语句的 dsql_id,其代码如下:

注意: 在查询的过程中使用 where 语句来限制显示输出的结果,以方便阅读。

```
SQL> select sql_id,sql_text
  2  from v$sql
  3  where sql_text like '% select e.lastname,%';
SQL_ID
------------------------------------------------------------
SQL_TEXT
------------------------------------------------------------
3ygg74tbfngnj
select SQL_ID,SQL_TEXT from v$SQL where SQL_TEXT like '% select e.lastname,%';
```

由查询结果可以发现,sql_id 为 3ygg74tbfngnj。

步骤 10 利用 sql_id 调用 dbms_xplay 软件包中的 display() 函数以显示刚刚执行过的 SQL 语句的执行计划。

```
SQL> COL PLAN_TABLE_OUTPUT FOR A100
SQL> SET LINESIZE 200
SQL> select PLAN_TABLE_OUTPUT
  2  from table(DBMS_XPLAN.DISPLAY_CURSOR('3ygg74tbfngnj'));
PLAN_TABLE_OUTPUT
------------------------------------------------------------
SQL_ID: 3ygg74tbfngnj, child number: 0
------------------------------------------------------------
```

```
select SQL_ID,SQL_TEXT from v$SQL where SQL_TEXT like '% select e.lastname,% ';
Plan hash value:903671040

----------------------------------------------------------------------------
Id    | Operation          | Name                 | Rows | Bytes | Cost(% CPU) |
----------------------------------------------------------------------------
0     | SELECT STATEMENT   |                      |      |       | 1(100)      |
PLAN_TABLE_OUTPUT
----------------------------------------------------------------------------
Id    | Operation          | Name                 | Rows | Bytes | Cost(% CPU) |
----------------------------------------------------------------------------
* 1   | FIXED TABLE FULL   | X$KGLCURSOR_CHILD    | 1    | 523   | 0(0)        |
----------------------------------------------------------------------------

Predicate Information (identified by operation id);
----------------------------------------------------------------------------
1 - filter("KGLNAOBJ"IS NOT NULL AND "KGLNAOBJ" LIKE '% select e.lastname,% '
AND "INST_ID" = USERENV('INSTANCE')))

已选择 20 行
```

由上面的结果可知,这两个表在连接时,最大的子表 employees 居然使用的是全表扫描(Fixed Table Full)。

15.6.2 缓存命中率实例

步骤 1 查询与缓存命中率相关的 3 个参数,其代码如下:

```
SQL> col name for a40
SQL> select statistic#, name, value
  2  from v$sysstat
  3  where name in ('physical reads','db block gets','consistent gets');

STATISTIC#  NAME                                     VALUE
----------  ----------------------------------------  --------
       159  db block gets                             89999
       163  consistent gets                           743277
       172  physical reads                            21243
```

步骤 2 利用公式计算缓存命中率,其代码如下:

```
SQL> select a.value + b.value - c.value "logical_reads", c.value "phys_reads",
  2  100 * ((a.value + b.value) - c.value)/(a.value + b.value) " buffer hit ratI/O"
  3  from v$sysstat a, v$sysstat b, v$sysstat c
  4  where a.statistic# = 159 and b.statistic# = 163 and c.statistic# = 172;

logical_reads   phys_reads   buffer hit ratI/O
```

```
         812105      21246          97.450534
```

步骤3　进行查询,结束后观察3个参数的变化,其代码如下:

```
SQL> select * from hr.employees;

已选择 107 行.

SQL> select a.value + b.value - c.value "logical_reads", c.value "phys_reads",
  2  100 * ((a.value + b.value) - c.value)/(a.value + b.value) " buffer hit ratI/0"
  3  from v$sysstat a, v$sysstat b, v$sysstat c
  4  where a.statistic# = 63 and b.statistic# = 67 and c.statistic# = 72;

logical_reads  phys_reads   buffer hit ratI/0
-------------  ----------   -----------------
   1165458361           0                 100
```

步骤4　第二次执行该查询,观察参数变化,其代码如下:

```
SQL> select * from hr.employees;

已选择 107 行.

SQL> select a.value + b.value - c.value "logical_reads", c.value "phys_reads",
  2  100 * ((a.value + b.value) - c.value)/(a.value + b.value) " buffer hit ratI/0"
  3  from v$sysstat a, v$sysstat b, v$sysstat c
  4  where a.statistic# = 63 and b.statistic# = 67 and c.statistic# = 72;

logical_reads  phys_reads   buffer hit ratI/0
-------------  ----------   -----------------
   1165458361           0                 100
```

可以看到,物理读次数没有发生变化,这是因为第三次执行该查询,所有数据都在内存中,不需要从外存中读取数据,即不需要物理读,因此物理读次数没变化;而逻辑度增加了多次,所以命中率又有所提高。

理论上反复执行该语句,命中率可以达到100%。

步骤5　作为一个DBA,监控各种内存的命中率是其职责之一。根据经验,要想保持Oracle服务器具有较高的性能,数据库块缓存的命中率应该大于90%,共享池库缓存命中率应该大于99%,共享池中字典缓存命中率应该大于85%。

共享池中库缓存应大于0.99,其代码如下:

```
SQL> select sum(pins - reloads)/sum(pins) from v$librarycache;

SUM(PINS - RELOADS)/SUM(PINS)
-----------------------------
                    .994198223
```

共享池中字典缓存应大于0.85,其代码如下:

```
SQL> select sum(gets - getmisses - usage - fixed)/sum(gets) from v$rowcache;

SUM(GETS - GETMISSES - USAGE - FIXED)/SUM(GETS)
-----------------------------------------------
                                     .955618225
```

图书资源支持

感谢您一直以来对清华版图书的支持和爱护。为了配合本书的使用,本书提供配套的资源,有需求的读者请扫描下方的"书圈"微信公众号二维码,在图书专区下载,也可以拨打电话或发送电子邮件咨询。

如果您在使用本书的过程中遇到了什么问题,或者有相关图书出版计划,也请您发邮件告诉我们,以便我们更好地为您服务。

我们的联系方式:

地　　址: 北京市海淀区双清路学研大厦 A 座 714

邮　　编: 100084

电　　话: 010-83470236　010-83470237

客服邮箱: 2301891038@qq.com

QQ: 2301891038(请写明您的单位和姓名)

资源下载: 关注公众号"书圈"下载配套资源。

书圈

获取最新书目

观看课程直播